The Recycle-to-L

Garden Composting

How Garden Recycling Works

Sustainability in your Garden Environment

Hands-on, know-how and understanding of practical delivery of integrated, closed loop gardening from waste, delivering real sustainability.

Plus - Safety Check List and How to calculate your own positive garden Carbon Footprint.

Bill Butterworth

First edition published in 2009
© Copyright 2009
Bill Butterworth

The right of Bill Butterworth to be identified as the author of this work has been asserted by him in accordance with the Copyright, Designs and Patents Act 1998.

All rights reserved. No reproduction, copy or transmission of this publication may be made without express prior written permission. No paragraph of this publication may be reproduced, copied or transmitted except with express prior written permission or in accordance with the provisions of the Copyright Act 1956 (as amended). Any person who commits any unauthorised act in relation to this publication may be liable to criminal prosecution and civil claims for damage.

Although every effort has been made to ensure the accuracy of the information contained in this book, as of the date of publication, nothing herein should be construed as giving advice. The opinions expressed herein are those of the author and not of MX Publishing.

Paperback ISBN 9781904312673
Published in the UK by MX Publishing
335 Princess Park Manor, Royal Drive, London, N11 3GX
www.mx-publishing.co.uk

This book is dedicated with great respect to gardeners, the men and women who have taken thousands of years to make gardening what it is and to help us all understand how to make it sustainable for another thousand years.

If you do nothing else with this book but read Chapters 1 and 11, you will find the "why do it" and "calculate the Carbon footprint for your garden". If that is enough, that's fine. If you want to know how it all works, that's the bit in between. If you do not have some basic school chemistry, you will find the language new but take it one step at a time, not too much at once, and you will get there. This is really exciting in terms of the contribution you are making to "being green" and understanding what everyone else talks about but most know little of how it really works. If you can manage to get through it and grasp most of it, then you will be able to answer a fundamental question about whether the way you garden is really green; if you carry on doing what you are doing now, will it all still be working just the same in a thousand years?

CONTENTS - (Pages in Brackets)

1. **Firstly Composting** *(6)*
   Why Do It and Why Don't Do It?
   Do it; Dangers; Soils

2. **The Closed Loop** *(15)*
   Hydroponics; How the Closed Loop Really Works; Why Natural Ecosystems Don't "Leak" Nutrients and Pollute Themselves

3. **Garden Composters** *(35)*

4. **Inputs** *(43)*
   The Garden Itself; Carbon, Green Wastes, Wood, MDF, Paper. Nitrogen Sources; Atmospheric, Urine, Green Material; Phosphorus, Potassium, Calcium, Sulphur and Trace Elements

5. **How to Manage the Nutrients in Waste** *(71)*
   Nutrients in Waste. C:N Ratio, Trace Elements and Nutrient Balance.

6. **The Structure of Soils** *(75)*
   In the Surface Soil; Sands, Clays, Loams, Chalk, Organic Matter, Fibres; The Soil Profile and Root Growth.

7. **Liquids and Compost Moisture** *(80)*

8. **Compost - The Finished Product** *(86)*
   Fertiliser, Soil Improver, Mulch

9. **Time and Soil Bio-Activity** *(90)*

10. **Human Health and Waste-to-Soil** *(92)*

11. Your Carbon Footprint *(97)*
   Does it Matter Anyway?
   The Concept of Carbon Footprint – How to Work Out Yours Within the Garden.
   How Your Garden Can Affect What Happens Outside the Garden.

APPENDIX

References and Further Reading (109)
Number Referencing shown throughout the book in subscript.

About the Author (113)

Chapter 1
Firstly Composting
Why Do It and Why Don't Do It

Why Do It?
There is something about "waste" and gardening which is what an academic would call "beautiful". One of the ways of solving any problem is to find an exact opposite and put the two together; both problems disappear. Many will have heard, on television or the radio, of a "black hole" which can occur in outer space. For those who know a little of what a "black hole" in space is, and indeed those who don't, the analogy of a black hole and a supernova is an example of this way of problem solving. Get the right size of black hole and the right size of supernova and put them together and there is nothing; the mass of the black hole and mass of the supernova cancel each other out. The down-to-earth gardening problem of what to do with untidy "wastes" and how to make things grow has a similarity; the wastes from the house and the garden itself is one problem and the need to produce flowers and vegetables which need nutrients to make them grow is the other. Put these two problems together in the right way and the problems not only disappear, they produce a benefit. So, what clues have we in discovering, unearthing maybe, what "the right way" really is?

The Living Soils Jigsaw
There is a jigsaw puzzle relating to "waste" which puts some pieces together to make the beautiful solution.

Jigsaw Piece No 1
To put it simply; "waste" is a problem in urban societies and growing plants in the garden needs fertiliser. What man did to make the waste is generally

to take things out of the earth and manufacture something. That manufactured thing or the process which created it will have some "waste" somewhere attached to it and that waste will have in it the "fertilisers" or nutrients which are needed to grow plants.

Jigsaw Piece No 2
Now, all of that is a pretty good idea because most farmed crops in the world use fertilisers made in a factory, commonly and usefully referred to as "mineral fertilisers". There are many elements which a plant needs in order to grow well. Some are in small amounts and we generally call these "trace" elements. Most mineral fertilisers are pure chemical and have little or no trace elements. There are three high-volume fertilisers; Nitrogen, phosphate and potash. The starting point for making Nitrogen fertiliser is to pass air through a two meter diameter electric arc. The temperature in the arc is enough to make the Nitrogen in the air (which is normally an inert gas) react with the Oxygen to form Nitrogen oxides and that is the first stage in making the Nitrogen fertiliser. It is important to ask where the electricity came from; in nearly all cases from burning fossilised fuel, of course.

Jigsaw Piece No 3
A plant is interesting in many ways but there are two characteristics worthy of note here. The first is the potential to use the components in "wastes" as nutrients; most of us can accept the idea of composting biodegradable wastes. The second involves a bit of chemistry but you really do need to grasp this and it isn't too difficult. Look for a moment at what the plant uses to make the vast bulk of its tissue; and it is not the nutrients in the waste.

A plant takes water in through its roots and Carbon dioxide in through its leaves to make large Carbon molecules which most of us call "organic matter". Even a non-chemist can follow the basic chemical equation which can be seen as:

$$6CO_2 + 6H_2O \quad \text{makes} \quad C_6H_{12}O_6 + 6O_2$$
(via the chlorophyll in the leaf)

Put in another way:

Carbon dioxide   (in the plant produce)   sugar
and water   and Oxygen

Note: a bit of chemistry and arithmetic: the number of C's, H's and O's on each side of the equation add up to the same.

That is a staggering equation. It is the one on which the whole of sustainable life on earth is based. Plants mop up Carbon dioxide and give out Oxygen. Animals (including humans) take up the Oxygen and, by burning sugar to release muscle energy, give Carbon dioxide back into the cycle. This is a cycle and this book will return to the idea that "cradle to grave" analysis means little; we need to think in sustainable cycles to make everything circulate round and round[2]. That is what gives true sustainability.

That equation and "the closed loop" is truly staggering. However, there is an implication which we are prone to forget.

Jigsaw Piece No 4
The Carboniferous Era in the history of the earth started some 360 million years ago. It lasted around 60

million years. In that time, enormous quantities of Carbon dioxide were taken out of the atmosphere using the above basic equation and turned into large Carbon molecules leaving us an air we could breathe. Those large Carbon molecules formed the crude oil, gas and coal reserves which we will burn, probably nearly completely, inside maybe 160 years. NASA, the American space agency, might forgive the observation that "Houston, we have a problem". When we burn the fossil fuels, the equation is reversed. When we burn the fuel, we use up Oxygen and give out Carbon dioxide, the greenhouse gas we talk of so often. Not only will we run out of fossilised fuels, where will the Oxygen have gone? "Global warming$_8$", which we now talk about daily, is not just about the earth getting a bit warmer. It will have a very different climate and there will be a little less Oxygen in the air we breathe.

Jigsaw Piece No 5
It was not until 1843 that John Lawes, generally viewed as the father of the world-wide, mineral fertiliser industry, built his first factory. Until well into the 1950's, most of the world made fertilisers for its crops from animal manures, soiled straw bedding and local industrial wastes. Mass production of mineral fertilisers did make, and still does make, a very real contribution to feeding people. That contribution is so big the idea of just abandoning mineral fertilisers is fanciful. However, just sticking with them is also madness. We do have to find better ways of producing food.

Jigsaw Piece No 6
The old boys and organic "cranks" were right; there really is less crop disease in organically fertilised gardens.

The Beautiful Conclusion
If we take urban wastes, world-wide, to your garden and farmers locally (locally means *saving no. 1* because transport costs energy) world-wide, then we can grow plants and crops in your garden with less, and in some cases no, mineral fertilisers. This means we would not then be burning fossilised fuels to make electricity to make the fertiliser (*saving no.2*). If we grow crops which produce oil seeds, then we can make biofuels and avoid burning fossilised fuels (*saving no.3*). The figures on Carbon dioxide savings are very large and enough to make a significant contribution to reducing global warming and putting Oxygen back into our atmosphere (see Chapter 11).

<u>*Why Don't Do It?*</u>
All of that seems a pretty good idea, and it is. However, there is a downside with both trivial and potentially serious consequences.

Trivial first. Recycling is hard work. It is messy and often looks it. It occupies space. It is much easier and tidier to just throw everything in the bin and then go and get a bag of mineral fertiliser from the garden centre. Alternatively, go and get some horse manure, or farm yard manure, or compost from a big local centralised site: they are all organic and, at least in some way, good alternatives.

<u>*Dangers*</u>
Potentially serious consequences? Chapter 3 will give the names of the really dangerous bugs from a human health point of view. Also, look at Figure 2.5 on page 31, at the section of the compost temperature curve. The third part of the process is mainly fungi and many of the fungi are actually Penicillins – not highly anti-biotic and not

dangerous either. However, the first part is mainly bacteria and they might be a different story – again, see Chapter 3 for a potentially lethal risk, but, in the meantime, remember that a compost heap is not a place for kids to play and everyone should wash thoroughly after tending a compost heap.

Frankly, there are disease dangers with compost, not very large ones if the heap is managed properly (which with small heaps is quite difficult). And good composting involves significant space, some knowledge and some hard work. It is a bit more than just making a heap of waste. (All that is, is a heap of waste!) You would do better to just dig a trench, only a spade depth, and put the wastes into it and cover it up. Better still, just dig it in and mix it up with soil. Looks a mess, works well, does not control weed seeds if they are present. Nevertheless, it is worth remembering that there is very little that can be done in a compost heap that cannot be done in the soil itself.

The really good news about the compost heap is that if there is a material which has, or might have, potentially dangerous organism which might be pathogenic to humans, then the compost heap is the place to put them. A compost heap will sterilise these pathogens$_{37}$. If you are already composting, study Chapter 3, and then look at the following Safety Check List to protect you, your children and your pets. If you are just starting composting, then try and get your head round the principles in this book and then come back to this check list; you won't go far wrong.

SAFETY CHECK LIST
1. Educate the children in a balanced way. Tell them how this is a small contribution to a better environment, because it is. Without scaring them or other people, tell them that there maybe some fairly nasty "bugs" in a compost heap and that these can cause bad tummy ache and diarrhoea, maybe worse.
2. Show them the garden thermometer and show it in use. They need to know about pasteurisation. (Just as in milk.)
3. Always use gloves when handling compost.
4. Even if you have used gloves, wash hands afterwards and especially before eating. Never put dirty hands near the mouth.
5. Keep it all tidy, best use a bin or wooden walls lined with carpet and do not leave food remains outside the compost bin.
6. Keep it moist, mix green and brown materials and add only small amounts to avoid overheating.
7. Look out for rats and mice. Control them safely.
8. If you choose to put materials direct to soil and avoid use of the composting process, it is okay to put garden remains on the surface of ground as a mulch, BUT, if you have kitchen food remains and you choose to put them direct to land, bury with at least 150mm of soil on top and avoid dogs or other animals being able to dig them up.
9. Keep it tidy. Watch the temperature. Wash before eating.

One more point about temperature; the remaining advantage of a compost heap is in weed seed kill and in a "crop window". If you have a full flower bed or vegetable patch you may need to wait for the end of the season and the gap before the next planting; in order to spread the compost.

Soils

The truth is that the old boys, the organic "cranks" with sandals and ponytails, were right (again! at least in part). We really do have evidence of why this is, what the mechanisms are and how to encourage and manage them.

Have you ever thought about why the tropical jungles do not pollute the river running through them? Take, for example, the Amazon River Basin in Brazil, South America. The trees in the rainforest grow 70m tall. It follows that there must be really good fertility in the soils with plenty of nutrients including Nitrogen. (Nitrogen is one of the major plant foods which the plant uses to make proteins which build its cell walls. Nitrogen is also part of "nitrates" which cause pollution of groundwater.) Now, that area has not only all that fertility and plant nutrients, including Nitrogen, it has rainfall about as high as anywhere. How is that managed without putting nitrates into the groundwater? Why isn't the Amazon River full of green slime and dead fish? Take another example: the Fens in the Eastern part of the UK. When the Dutch engineer, Vermoyden, was employed to drain them around 300 years ago, some of them were 10 to 15m deep with black, organic soils. It was possible to grow crops, very good crops, and export the harvested product (and all the nutrients including the Nitrogen contained in plant proteins) without ever adding any sort of fertiliser. The nutrient reserves, including

Nitrogen, were enormous. How is that managed without putting nitrates into the groundwater? Why weren't the dykes and the Norfolk Broads full of green slime and dead fish?

The answer is that all natural ecosystems work on what everyone calls the "Closed Loop". If we know how that works and can imitate it, then we can garden and farm, we can sustain the human race, without destroying the long-term survival of our planet - if we can? We can and you can do it in your garden. We now know what that mechanism is and have at least a start on how to manage it and sustain it.

Chapter 2
The Closed Loop
Hydroponics, How the Closed Loop Really Works, Why Natural Ecosystems Don't "Leak" Nutrients and Pollute Themselves

Hydroponics
It is possible to grow plants hydroponically, i.e. in water with nutrients added. It has been around for centuries. It is still around. It is raised every so often as a way of feeding the world. Millions of US Dollars, GB Pounds Sterling (whatever currency) have been poured into it. Mostly, it does not work very well. It easily breaks down with plant disease outbreaks.

It works better if there is some solid material, say Vermiculite (which is an expanded volcanic clay). The solid material gets dirty, builds up (usually) a green algae; then it works better. Keep adding the dirt and it works still better. Keep adding and you will get to soils. Mankind has rather a lot of experience of working with soils and we can, on the whole, manage them rather well.

How "The Closed Loop" Really Works
The processing capability of the soil dwarfs human industry. An American scientist once calculated that the micro-organisms in an acre of arable soil would weigh as much as a fully grown cow[10]. In a gram of dry soil[1], there will be of the order of 100 million bacteria weighing one and half *tonnes* in a hectare (2.4 acres). The fungi will weigh maybe 2 tonnes in a hectare, even 3 or 4 tonnes in a really good organic soil with plenty of compost and biological activity. Actinomycetes, algae, protozoa and nematodes will add maybe another 2 to 3 tonnes per hectare. Earthworms will vary widely but may well be

another tonne per hectare. Add all the other invertebrates and there may be approaching 10 tonnes per hectare of small creatures, all full of nutrients, multiplying and dying at enormous rates, building up the black, tarry material we call "humus". In a compost heap, the figures are higher, often much higher. That living and dying has enormous activity attached to it; enormous processing activity to break up almost anything added to the soil.

We are talking tiny organisms of nano-scale and less ("nano" means 10 to the minus 9, eg, one nanometre is one $100,000,000^{th}$ of a metre). This is not just millions, not billions, but trillions of micro-organisms in just a handful of soil, a dynamic universe with enormous processing ability. Multiplication rates and biodiversity are enormous by human standards, and so is the range of their appetites. It is generally true that if any material, *any* material, is spread out far enough and there is enough time given, nature will deal with anything and bring the system back to a dynamic, balanced "normality" (where "normal" means sustainable).

Stop for a moment and consider the multiplication rates of some of these organisms. Take an example of a small one which you can see - a greenfly (the aphid which carries "blight" to your potatoes and many of your garden plants). The green fly can multiply asexually - one can produce without a mate. Take one and give it ideal conditions and say none die. Then one can cover the entire earth, including the oceans, 100mm deep in 24 hours. The soil micro-organisms are similar, but more so. These numbers make the numbers of the 2009 world financial crisis look small! Now look at how nature harnesses these organisms in the closed loop and how your garden really grows.

Why Natural Ecosystems Do Not "Leak" Nutrients and Pollute Themselves

Note; As a credit to original source and to add credibility to what is written below, much of the material in this section on nitrate pollution was first published by the author of this book in papers published in "Resource", Journal of the American Society of Agricultural and Biological Engineers in April 2001$_{3\ and\ 11}$ and in "Landwards", Journal of the British Institution of Agricultural Engineers in Early Summer issue 2002$_7$ and is reproduced here by kind permission of those original publishers. The original papers were published for a farming and engineering audience but the principles apply everywhere, including your garden.

Understanding the mechanisms in what is commonly called "The Closed Loop" and managing those mechanisms makes recycling to land dramatically safer in environmental terms. The figures below show the principles of The Closed Loop. Organic materials, and inorganic ones which have food value for the soil micro-organisms, do NOT breakdown directly to form "humus". Such materials added as "waste" are eaten by micro-organisms and turned into their own bodies. It is the breakdown of these bodies which form the stable black tarry material which gives soils their dark colour and which we call "humus". So, knowing how to feed these organisms is the first step in the management of composting and the soil. It is also important to see that the compost heap and the soil are not separate operations. Mostly, everything that goes on in a compost heap would also go on in the soil, even pathogen destruction. The big advantage of composting is to use the temperature for weed seed kill. What the micro-organisms do is feed, multiply and die – at enormous rates. Remember, it is the dead bodies which breakdown to form the black humus tar. Humus is a very, very complex mixture of heavy molecules of hydrocarbons (yes, this is the same process which makes crude oil), plus carbohydrates and proteins (which lock up the Nitrogen). These molecules are large and insoluble. There is no limit to the quantities that can be put onto the soil safely. The evidence for that is any natural ecosystem; but take, for example, the Fens again. When Vermoyden drained them, some were 10 to 15m deep. Farmers could grow crops every year from then and until now, exporting the harvested products with the nutrients they contained, including the Nitrogen, and never need to add any fertiliser. That was because the nutrients were locked up in big molecules which were insoluble in rainwater, so they never leached out into the groundwater. But how do they get into the plant? It cannot be in solution because if they were, they'd leach out. So how does it happen?

Those large molecules will sit there forever until long strands of soil fungi called mycorrhiza (pronounced my-cor-iza) eat them at one end while the other end of the fungal strand has a very close relationship with a plant which calls for food. These mycorrhiza either go up to and round the plant root hair, rather like the placenta in a baby mammal in the womb, or actually cross the root hair wall into the plant. This is a closed pipe or "conduit" and that is why the system does not leak! As common sense might indicate the plant and the fungi evolved together over millions of years and they operate at the same soil temperatures and conditions, so when the plant is hungry, the fungus feeds it; the system is demand-led. This system is how all natural ecosystems not only eliminate nitrate pollution, they eliminate all such out-of-balance pollution including phosphates, potash, etc$_{14}$. There is a further advantage: the system locks up carbon in the soil. The 100 million tonnes of "waste" produced in the UK and which could be recycled to land would, if incinerated, produce around 75 million tonnes of Carbon dioxide per annum which is 10% of the Kyoto Protocol estimate of total UK emissions. Composting to land can lock that up.

The following diagrams show how natural ecosystems manage to "leak" enough, and only enough, to keep the system working without pollution or starvation, i.e. in balance or "sustainably". Figure 2.1 shows a conventional view of how the system works and this is still taught in many colleges and universities round the world.

Just pause for a moment to absorb a bit of chemistry language. When a mineral fertiliser material goes into solution, ammonium nitrate, for example, it splits up into two small particles; the ammonium bit and the nitrate bit. These bits carry an electrical charge and they are called "ions". The ammonium bit has a positive charge (and is called a "cation") and the nitrate has a negative charge (and is called an "anion"). Sands will hold not very much of either ion, so the nitrate fertiliser, which is very soluble, will leach out in the rain very easily and quickly. You will loose maybe over 50% of what you put onto your soils this way. Clays will not hold much anion (the nitrate) but they will hang onto much of the cation, the ammonium which has a chemical formula of NH_4 – that means it has four Hydrogen atoms in its molecule and one Nitrogen. Fortunately, the soil system can use that Nitrogen to make proteins in the plant. So, this does work but the losses are high (i.e. much of the nitrate is lost). Also, not everyone has some clay. Good "loams" have some sand (so that they can "breathe") and some clay (which will absorb and hold onto nutrients).

Now see how this works in the diagram and see the logic of how losses occur with mineral or "artificial" fertilisers. Those losses are not only financial, they also pollute the groundwater.

Figure 2.1
The conventional view of how plants feed with the assumption that nutrients get to the plant via solution in the groundwater. With mineral fertilisers, this probably is true either partly or completely.

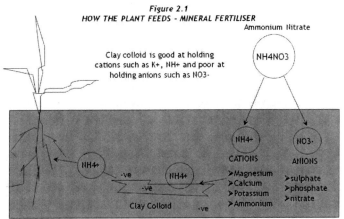

Figure 2.1
HOW THE PLANT FEEDS - MINERAL FERTILISER

When mineral fertilisers such as Ammonium nitrate are applied, the cations are held in the soil colloid "bank" which also holds water. However, rain will take nearly half of the nitrate into groundwater.

This description is not wrong, especially when referring to agriculture which uses mineral fertiliser. However, it is incomplete and potentially misleading, especially if it were applied to soils with significant levels of humus or organic matter. Further, it does not explain why natural ecosystems don't leak enough to cause pollution. Figure 2.2 shows how such pollution is avoided and shows the mycorrhizal conduit which is the central mechanism in what is commonly referred to as "The Closed Loop". It is the mechanism which stops leakage at a level of pollution. It is this mechanism which feeds plants and protects them from disease[17].

Before studying the diagram, it is important to note that "organic matter" may contain any materials derived from living organisms. However, "humus" is a black tar which gives soils their black colour and has been produced by micro-organisms digesting organic matter and then dying. The black tar, then, is the juicy remnants of the dead bodies of micro-organisms.

Figure 2.2
In natural ecosystems, plant nutrients do not enter into solution in the groundwater in order to enter the plant. Humus is a complex mixture of heavy molecules which are not soluble in water. Neal Kinsey, in his book "Hands-on Agronomy[14]" points out that this humus has several times the sponge or "colloidal" capacity of clay and will hold onto anions as well as cations. That, however, still did not explain how the nutrients got into the plant without leakage. It was the American PGA (Professional Golfers Association) in the USA who pursued this investigation to show that the soil fungi, known as mycorrhiza fed, at one end of their hyphae, on the humus, and the other end went not up to somewhere near the plant root hair but across the root hair wall into the plant. This finding was added to by researchers at Aberystwyth in South Wales who showed that there was another type of mycorrhiza which went up to the root hair and wrapped around it much as the placenta in a mammal. This is a molecular level relationship. This is a closed conduit. That is why the natural ecosystems do not leak.

Figure 2.2
NATURAL ECO-SYSTEM

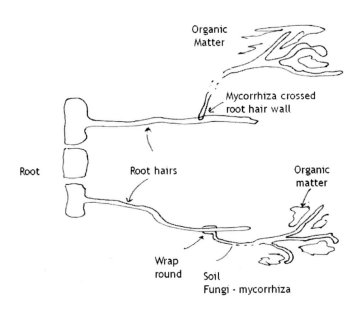

Figure 2.3
As Figure 2.2 but more formally.

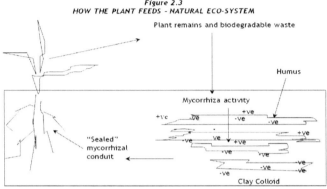

Figure 2.3
HOW THE PLANT FEEDS - NATURAL ECO-SYSTEM

Mycorrhiza are the key to pollution control because they give a "Closed Loop" to recycling both cations <u>and</u> anions.

Figure 2.4
What composting can do is provide a "buffer" between a controlled process and the soil. That buffer can isolate physical, chemical and biological risks in order to allow processing, monitoring and safety controls to operate[1].

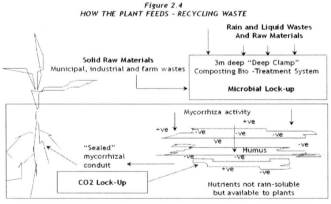

Figure 2.4
HOW THE PLANT FEEDS - RECYCLING WASTE

How the closed loop gives pollution control and scope for treatment systems.

Why Soil Universes do not Pollute Themselves
It has been said that there are more stars in the sky than grains of sand on the earth. Think, then, of the soil as a universe with unbelievable numbers of particles and organisms all running round doing their own thing. Soils manage pollution in two ways. It is, perhaps, useful in understanding how these mechanisms work by first understanding what pollution is. Pollution is, incidentally, not just quite natural but fundamental to life itself. Part of the definition of a living organism (as distinct from not living – a tractor, a computer or even a plastic duck for example) is that the living organism is continually producing wastes which would be toxic if they stayed inside the body; we call these materials "pollutants". These pollutants are products of processes going on inside the body of the organism and these products will become increasingly toxic (because of the volume of them) as they build up and must be got rid of, outside that body. The question then arises as to when that production of pollutants becomes "pollution". In the scientific sense: always. In the practical or legal sense: well, it depends. If the production of pollutants is at a level where the local environment cannot, in time but fairly quickly, bring the system back to what the ecosystem previously was, if there is a shift in the ecological equilibrium, then, it may be said, pollution may have occurred. The difficulty in any discussion about being definitive is that nature, if you give it long enough, will solve just about any pollution. (Fortunate really, isn't it?) In any case, all real life situations are what a scientist calls "dynamic" - they are living and changing.

The two ways a soil combats this pollution are by providing a "buffer" to buy time and by digesting the pollutants and passing them into the ecological chain. A compost heap is a man-controlled buffer which allows time for the process within the heap to break the toxins.

Soils, which are substantially sands, have little buffering capacity and little ability to hold mechanically onto any particles – large or small. For example, put ammonium nitrate fertiliser onto a sandy soil and the ammonium cation and the nitrate anion will leach out very easily with probably more than half of the Nitrogen plant food going into the groundwater with rain or irrigation. That is a significant economic loss and potential pollution of groundwater. However, put the same material onto a clay soil and the colloidal capacity of the clay will hold onto much of the ammonium cation and, maybe, a little of the nitrate anion, too.

Now, having understood that, there is a staggering quality of humus which makes everything that the old organic enthusiasts advocated potentially true; humus is different. That difference is that humus can hold onto otherwise soluble materials which would leach out in rain or irrigation. Put ammonium nitrate, for example, onto humus and, if there is enough humus, there will be a retention of all of both the ammonium bit and the nitrate bit – pretty nearly all of both ions!

There will be no leaching with rain or irrigation of either the ammonium or the nitrate ion. Pollution of groundwater will be eliminated. So, different soils will have different buffering effects and we can alter that capacity by adding and managing the organic matter levels of soils, specifically the humus content.

The second mechanism for controlling pollution is to breakdown the pollutants. As the PGA golfers in the USA found out, their golf greens of sands had little or no humus and, therefore, little biological activity with very low populations of mycorrhiza. We have known for some time of the anti-biotic activity of many of the soil mycorrhiza[13 and 17]. Many are, incidentally, Penicillins. If those micro-organisms are there, then they can and do breakdown almost anything given enough time and dispersion. Here, again, is why the original organic gardeners were on the right track. Maybe they weren't so "cranky" after all.

One more staggering thing about mycorrhiza. There is some evidence that when mycorrhiza are living in a soil and feeding the plants in that soil, and a herbicide is sprayed on the plants, one of the reasons why the herbicide works involves the mycorrhiza sensing the activity of the spray and extracting food supply from the weeds and feeding the crop[37]. It appears that the mycorrhiza can sense the stronger partner and drag food from the weaker plants to feed the stronger. This might be expected in terms of evolutionary survival of the fittest. There is, however, a cautionary note; if that mechanism is there, then putting the spray on in a way that weakens the crop garden plant before or more than the weed and, guess what, the mycorrhiza will weaken the garden plant. This would explain some previously misunderstood failures of spraying. We have much to learn about managing the soil mycorrhiza but there is a basic rule; if the natural mechanisms of nature are to be interfered with, then we'd better be careful and get it right.

On the subject of the mechanisms of the compost heap and what materials they will breakdown, a relevant scientific paper[18] was published at the British Crop Protection Council Symposium, November 2001. S.C. Rose and colleagues delivered a paper on "The design of a pesticide and washdown facility". What basically happened in this research by ADAS (a private research company and commissioned by Defra (HM Government's Department for Food and Rural Affairs) was concerned with examining what happened when plant protection spray chemical, either as undiluted active ingredient splashes or as diluted washings, were released into the environment and how any pollution risk could be controlled. The experiment took a cocktail of some fairly dangerous active ingredients and put them onto a number of surfaces ranging from concrete, through gravel and sand and to a "bio-bed" of straw, loam, non-peat compost with grass planted on top. All except the bio-bed were almost without effect; the bio-bed was almost totally effective in breaking down the cocktail and eliminating pollution risk. That bio-bed was half a metre deep and presented ambient temperatures. That bio-bed could be bettered by your compost heap, provided it is biologically active and achieves a temperature in the 50 to 90°C range. We shall come back to how to secure that activity later.

This leads us to a conclusion and, of course, naturally, to a further question. Firstly, yes we can manage these mechanisms. We can introduce a buffer (such as a compost heap in a controlled and "imprisoned" situation). We can provide conditions which will encourage and manage the biological activity which will destroy pathogens and undesirable toxic molecules[8 and 15]. The remaining question, however, is how to identify and manage the need for enough time and enough dilution to allow these desirable functions to reach an identifiable and acceptable end point within a predictable time frame. Spreading things out far enough and allowing enough time develops in any gardener in time as a matter of what people sometimes call "instinct". If you need a more structured way to be right first time, the scientists call it "Dispersion Technology" which is a posh name for careful observation, intelligent common sense action, and a more formal way of disciplining the process so it can be repeated.

Odour, Health and Safety - Controlling the Process

John Lawes, generally regarded as the "father" of the mineral fertiliser industry, did not start his first superphosphate factory till 1843. Progress was initially slow but "artificial" fertilisers eventually became cheap to produce, very concentrated (so were easy to transport and easy to spread), and they made dramatic and visual differences to crop appearance and yields. Once large-scale production became available, it took only 15 to 20 years to sweep the world. These artificial, or mineral, fertilisers are a comparatively recent invention. Until they arrived, farmers used mainly locally available wastes, but the great sailing ships also transported guano (bird and bat "droppings" or faeces) across the oceans for this same purpose of fertilising crops. When locally available wastes, such as cotton "shoddy", leather trimmings, slaughterhouse wastes, were available, they were used as fertiliser. Grain was imported from overseas, fed to livestock and the urine, dung and soiled bedding "muck" was used for the same purpose. So was sewage. Waste has always been recycled to land[5 and 37], ever since *homo sapiens* began to wander and even before he settled. Regulators in most countries in the world often fail to understand that everything we have came from the land and it will eventually go back there. If it is spread out far enough and there is enough time, nature will have its way.

The basic process of composting, or any other aerobic biological process, involves four needs; a feedstock with reasonably balanced nutrients, an appropriate population of micro-organism species, moisture (water), and Oxygen$_5$. Generally speaking, if the material is of plant or animal tissues, it will have enough nutrients to operate reasonably normally in a compost process. In most environments, there are plenty of micro-organisms and of a wide enough range of species to operate the process. You really do not have to add "cultures" of special micro-organisms; they are there already. If there is not enough Oxygen, the process will go anaerobic and give off a bad odour and the process will slow down. If there is not enough water, the material may well go dark in colour but the process will be incomplete and will start up again when the material is incorporated into the soil after spreading. This may not matter unless there is a shortage of Nitrogen and, in such a case, the micro-organisms in the soil may preferentially use the easily available Nitrogen in the soil to build their own bodies in order to attack the energy source (the Carbon) in the added compost. Gardeners will sometimes call this "Nitrogen starvation". The Nitrogen is not lost and will be available to the garden plants at a later date when the biological activity of the soil catches up with balancing the Nitrogen status of different fractions of the soil.

This was described in a composite form in "The Straw Manual" by Bill Butterworth[20]. Previous to this date, farmers in the UK faced a ban on burning straw behind the harvester and it was necessary to find out how farmers could incorporate unwanted straw into the soil without high costs and/or loss of yields. That book collated the available research from world-wide sources and showed that the soil will live quite well with what might appear to be enormous imbalances (as with the high Carbon content of cereal straw) provided it is given a little help and time. The basic rule when changing to a system which involves putting large amounts of Carbon into a soil is to add enough mineral Nitrogen fertiliser in the first year to allow the soil micro-organisms to build the protein of their own bodies, allow a little less Nitrogen in the second year and less or none by the third or fourth year. After that, the soil system will cope because the added Nitrogen does not leach out and the soil micro-organism population has changed in species and population numbers to give the biological activity required to deal with the new regime.

A compost process is basically the same. It is worth going back to the basics of composting for a moment; the four needs being feedstock, micro-organisms, moisture and Oxygen. It might help to add "time" at this point. Now, if one of these basics is either not there, or not in the right quantity, then the process will change or slow down. For example, if there is not enough air, the process will slow down and either stop or go anaerobic which will give off a bad smell. If there is no change of gases, the process will eventually stop. Similarly for water; lack of it will slow the process eventually to the point of cessation. Exactly the same applies to the other two inputs of feedstock (obviously) and micro-organisms (less obvious but quite interesting).

An idealistic view of a composting process graph of temperature against time looks as follows. In practice, this will only occur in a compost of uniform material, uniformly shredded and uniformly aerated. Most actual operations will produce a patchier picture although of the same progression.

Figure 2.5

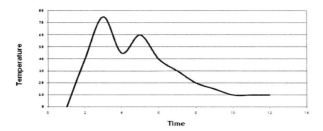

Figure 2.5
COMPOST TEMPERATURE CURVE

The first peak in temperature is mainly of bacterial activity. This is the risky time in terms of lack of Oxygen and odour production. Deliberate Oxygen starvation can and does produce Methane gas which can be used in heat and power generation. The top of this curve will normally be targeted at 55 to 65°C. That is the part of the process which will give you safety (see the section on "Composters".) Above this temperature gives increasing losses of Nitrogen (which is, of course, valuable as a fertiliser) and increasing risk of fire. In a garden compost heap, this is very unlikely (I've never heard of spontaneous combustion in a garden heap although it has been known to happen in large industrial scale operations). What, however, can happen is a very significant loss of not just the Nitrogen but of Carbon by oxidation and that means an overall and significant loss of humus. The second peak is of fungal activity; mainly "Pin Moulds", which are Penicillins. The trough in the middle is mainly of Actinomycetes, the fungi which give woodland its pleasant smell after rainfall. The curve on the right falls, but never actually to ambient unless the material, or the ground onto which the compost is spread, is frozen solid$_{23}$.

On odour risk; the first peak is the biggest danger period. If the heap goes short of Oxygen (it is then "anaerobic") it is likely to smell unpleasantly. The middle trough is not likely to cause offence but may be detectable. The fall away from the second peak is likely to be detectable and is a musty smell but not really offensive. Any smell is, of course, worst at turning. So, it is the turning during the first peak which you need to be careful about. Making sure the wind is in the right direction may be a sensible precaution if there are sensitive neighbours.

Following the logic of that curve and the four basic inputs into the process, it becomes clear that it is possible to influence the direction and speed of the process. It is possible to speed the process up and to slow it down but never make it fast. It is possible to make the process more odorous or less odorous but never without odour. It is possible to make the output look dark in colour and friable, but that does not necessarily mean that the process is complete and all the soluble nutrients have been incorporated into the bodies of micro-organisms, turned into humus and made safe to spread without pollution risk. Thoroughness can be speeded up but never made fast in composting. This is a biological process and time is fundamental to that process.

There is one other point to bear in mind. In most gardens, most of the time, the composting process is not in a single batch. A little bit is added now and a bit later. So, the process is continually changed by the addition of new material. So the curve starts off again and there are bits in the heap at all stages, all the time. However, there comes a point at which the process needs to stop and be allowed to complete. So a second heap becomes necessary and many gardens, if big enough, will have at least 2 and maybe 3 or 4 heaps. If space is limited to one heap or bin, then there are compromises to be made and some material is likely to be used immature. Unless weed seed kill is important, then some immature compost in selected places is not a major loss or difficulty in most situations.

Alternative Methods
If differentiating by detail, there are almost as many methods of composting as there are operators. What really matters is getting the compost up to at least 50°C, preferably 55° to 60°. Why that temperature? Well, Louis Pasteur, back in the 1860s, developed what we now call "Pasteurisation" which means that if most human pathogens (the "bugs" that cause our diseases) are cooked at 60°C, they die[16]. That magic figure of 60°C is now enshrined in Law. For example, if you were challenged in court for allowing other people's children to play in your compost heap and they contracted a disease from it, and you could show that the heap had risen to 60°C, then you could probably rebuff that challenge, or at least show you had taken reasonable care. In fact, we can do a little better than that. Professor Lynne Frostick's team at the University of Hull, showed that the bugs in a compost heap give best anti-biotic activity between 55 and 60°C[5], a little less, but it takes longer (which is okay by most gardeners).

There is a further major plus of well managed composting in terms of human and environmental safety when recycling wastes - especially imported wastes - where it may not be possible to be quite sure what is in them. A very large study was done by the Environmental Protection Agency in the USA (USEPA) in 1998 which came to the conclusion that composting will breakdown a very wide range of toxic chemicals into quite safe ones. These, the USEPA concluded[8], included gasoline, diesel fuel, jet fuel, oil, grease, wood preservatives, PCB's (Polychlorinated Biphenyl toxins), TNT and other explosives. We have to conclude that composting processes are fairly capable.

So, the temperature matters from a disease risk point of view and the break-down of potentially toxic material. It also matters from a weed seed destruction point of view; very few weeds will survive a well run composting process. That goes for living weeds, including couch grass and Japanese Knot Weed, too. But there is something more. Go back, for a minute, to the ideas on the closed loop and the temperature curve (figure 2.5). If the micro-organism activity is not there, then the soluble materials are not digested and will leach out. If they are not digested, it will take longer for the soil fungi (the mycorrhiza again) to feed on the materials and feed the plant. If the easily digestible parts of the "wastes" you put into the compost heap are not turned into humus, then they cannot be fed directly and by a fast route on demand by the soil fungi direct to the plant when the plant asks for food. So that activity can be measured by temperature and the 50° to 60°, maybe 65°C is a good guide. Above this and it is likely to lose Nitrogen by the micro-organisms turning the Nitrogen into ammonia which is lost to the atmosphere. Less than 50°C and the process will be unlikely to be complete unless it has been allowed a very long time, maybe several years. If the process is incomplete, then the Nitrogen may not be completely secure and may leach out with rainfall or irrigation.

Chapter 3
Garden Composters

In most circumstances, it is certainly worth considering just leaving uncomposted wastes on the surface of the soil as a mulch; the worms will take it down. Most people don't like this; it looks untidy and does not kill the weeds.

Figure 3.1

Figure 3.1
DIRECT TO SOIL

Looks untidy. Does not kill weed seeds unless buried deep. Otherwise works well.

In any case, digging has several negatives; it helps make weeds chit and grow, it accelerates the oxidation and loss of organic matter and it is hard work. If leaving uncomposted wastes on the surface is unbearably unsightly, you haven't the space, or whatever other reason which makes old habits die hard, then get composting.

A few precautions are observed by the sensible. By all means show the kids what is happening and help them understand that this is not a plaything. There is a risk, very small, of disease; most likely limited to tummy upset and diarrhoea but there are potentially more serious diseases. Salmonella is a likely cause of tummy ache but it could have more serious consequences. Some materials, especially animal faeces but potentially almost anything, anywhere, may carry a bug called Escherichia coli, "E. coli" for short. Nearly all strains are inoffensive or might cause a tummy ache. There is one strain, sometimes mentioned on the national news, with the label "Strain 0157:H7", and that can be very serious, even fatal. Animal faeces sometimes carry serious diseases; maybe Leptospirosis and Typhoid. Composting to the 55 to 60° *will* kill these organisms. To get a good kill, hold that range of temperature for 3 or 4 days at least, turn the edges of the compost (which will not have been so hot) into the middle and get the temperature up again and hold it for another 3 or 4 days. So do be confident of safety, don't get all screwed up about the dangers, *BUT* do work the compost and get the temperature up to get that safety. Until you get the hang of things, use a thermometer!

Also, watch out for rats; trap and bait sensibly – both traps and baits are potentially dangerous to wildlife and the kids. Bait laid at least 150mm up an old plastic drainpipe will protect it from most things other than the rats.

Figure 3.2

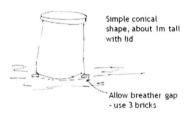

Figure 3.2
Your Composter

Simple conical shape, about 1m tall with lid

Allow breather gap - use 3 bricks

To turn, lift off the whole cone, place alongside and re-fill.

If you want a conventional plastic composting bin, they are not very expensive and you may be able to get a subsidised one from your local authority. The simplest is a black or green plastic cone, maybe a metre high with a plastic lid. These are really "tidiers", rather than a technological wonder. Remember the basics of composting; food, bugs, moisture and Oxygen. Well, a solid plastic wall may help keep the heat in but it also keeps the Oxygen out. These bins are filled from the top and are difficult to agitate and aerate. So, to make one work, jack up the base to let some air in. To help, some of the bins available have a few holes in the sides. Alternatively, and in any case, pull off the plastic cone regularly, put it alongside, and refill it with the same material. That will mix it up and let the air in. How often? Well it depends. If the compost is working well, it will heat up within a few days, as soon as the temperature begins to fall, turn it and that will speed it up again. What temperature? Keep a garden thermometer and aim at 55 to 60 but 50 to 65°C will do fine. If you don't need the compost for several months, slow down and don't turn so often.

Figure 3.3

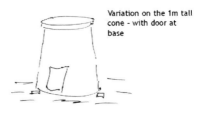

Figure 3.3
Alternative Composter with "Door"

In theory, the door allows removal of finished compost.

For a little more money, you can have a bin with a door at the base. In theory, this allows finished compost to be raked out at the bottom and allow a "first in, first out" continuous process. It is a bit difficult to get a full sweep of the base and there tends to be pockets left at each side in the base of the compost. It does not really matter but this does reduce the compost processing volume inside the bin.

There is one other thing to watch with these small composters; temperature. It is very easy for the heat generated by the micro-organisms to dissipate out and the process not only slows down but changes. At lower temperatures, the micro-organisms are of different species, they may cause odour and they will not kill the diseases which may affect plants in the garden, or you and the family. Low temperatures will not kill the weed seeds either. So get the temperature up; turn more and, if there is not enough Nitrogen present, add lawn mowings or other fresh green leaves, or add mineral Nitrogen. Best of all, do what the old senior gardeners in the Victorian era used to do: encourage the apprentices to pee on the heap. Joseph Jenkins, in his excellent book$_{40}$ records that one estimate of the value of urine is that it contains enough nutrients to grow food for that same person on a continuous basis. Urine has uric acid in it and that has Nitrogen in it. The children might like the idea, too. But not too much and you will need to turn it to avoid odour. (Make sure there are some sensible rules, appropriately enforced). The right bugs are there but you may need to encourage them. Use a thermometer. It is also very easy for the process to run out of Oxygen, especially if there are plenty of lawn mowings. Again, turning is the answer.

Want to do better? Well the bigger the heap, the more likely it is to warm up. Also, think about the balance between holding the heat in and letting the air in. A good and easy plan is to get four wooden pallets (a local builders merchants may be happy to let you have some for free) and line them, sides and top, with old carpet.

Figure 3.4

Figure 3.4
"Open" Composter

Pallets are sometimes "one way" and nobody wants them.

Four together make a really good compost surround. Line the sides and top with old carpet. Allow the front panel to detach.

You really need two areas, at least; one batch loading and one batch turning and maturing. The carpet will keep it tidy, keep it warm and let the air in and out. There is a small amount of Carbon dioxide produced in the compost process and a small amount of Oxygen consumed; the micro-organisms have life-processes much like us. However, it is important to note that these small amounts are dramatically recovered by many times the reverse when the compost is used in the garden instead of mineral fertilisers. Plants remove Carbon dioxide out of the atmosphere and pump Oxygen back in.

If you have a very large garden, lots of wastes and some free space out of the way, just make a big heap. If you keep it tidy, it will look like you know what you are doing and there will be less raised eyebrows. The unprocessed material at the edges will also be less and you will need to turn it less.

Figure 3.5

**Figure 3.5
Composting in Large Gardens**

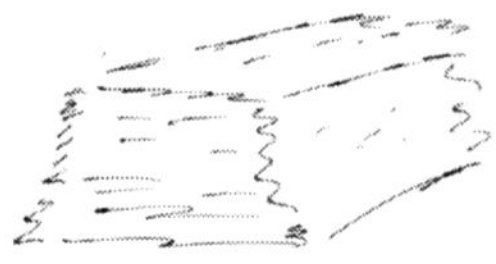

If you have a large garden with lots of space .....

..... Just make a big heap. Keep it tidy.

Shredding

There is no need to shred everything. Indeed, no shredding at all will not stop the process, but it may take longer. There is no need to break up or shred lawn mowings - to the contrary, they need mixing with some larger particles to let some air in as they will very easily slump into an anaerobic (without air) mass and smell. So, these small particles need mixing with larger, less easy to breakdown, materials so as to allow air retention - this is an aerobic (needs air) process. Where necessary, breaking up material, especially woody stalks, twigs and branches, before it goes into the compost process will certainly speed up the process. It does this for two reasons; there is a greater surface area of the material that has been broken or shredded (which allows the micro-organisms to get working on it), and shredding allows mixing with other materials (which allows the micro-organisms to get a balanced diet).

Small shredders, usually electrically driven can be purchased and are quite useful if they are not overloaded. Generally, it is things like hedge cuttings and woody prunings which really need shredding and that is especially true of Cupressus clippings.

Chapter 4
Inputs - Sources of Waste
The Garden Itself; Carbon, Green Wastes, Wood, MDF, Paper. Nitrogen Sources; Atmospheric, Urine, Green Material; Phosphorus, Potassium, Calcium, Sulphur and Trace Elements

Currently, British gardeners use millions of pounds worth of mineral fertilisers, with the basic raw material almost entirely imported. Most of it is sold in little packets at maybe up to a hundred times what farmers would pay for the same basic ingredients. What the gardener pays for is packaging, marketing, distribution, shelf space, retail staff; far too many fingers in the pie. All sorts of "wastes" could, technically, replace most or all of that. Could it be done safely and economically? Would recycling to land be of lower cost and more sustainable than current separation and collection routes? We may have become too obsessed with sophistication and centralised industrial processes when the most sustainable route has been in use for centuries[13]. The old boy in the country cottage garden; he often lived to be over a hundred years old and his garden soil was black with organic materials. He never threw anything away (even what we politely call "night soil" because there was no indoor lavatory). He recycled everything he did not need to the soil in his garden. He got all the immune stimulation he needed, at all the right levels, all the minerals and all the vitamins. Mostly, it worked incredibly well.

The Garden Itself
On the face of it, and logically, a garden ought to be able to support itself without additional fertiliser input. If the plant matter is all composted down during or at the end of the year, then it is all just going round again.

All of it, the trace elements, the Magnesium, the Sulphur, Calcium (lime), the potash, the phosphates and even the Nitrogen. This is even more likely to be true because, in the UK (more in tropical, high rainfall parts of the world) every hectare or acre of land gets about 2kg of Nitrogen fertiliser out of the rain every year. This comes from lightening flashes in thunderstorms – most of which you will never hear. The searing temperatures of a lightning flash are high enough to force the Nitrogen gas in the air (that Nitrogen is normally inert and will not react with anything) to react with the Oxygen in the air, forming nitric and nitrous oxides. These dissolve in the water in rain to nitric and nitrous acids. These are, of course, very dilute as acids but the Nitrogen is in the form of a fertiliser which the plants in your garden can use.

So, there should be at least a maintenance of all these nutrients and a build up of humus. Quite often, that is true; it does work.

However, you may remove significant quantities of these nutrients in cut flowers, vegetable crops and plants given away. Further, and more insidiously, some of the nutrients, particularly Nitrogen, will leach out in the rain. Not only that, the more you cultivate the soil, the more the organic matter will oxidise and go off into the atmosphere or groundwater. This is particularly true of the Nitrogen in the soil. Nitrate fertilisers will leach out in the rain or irrigation very easily, even on a clay soil, more so on sand. Typically, you will lose nearly half the fertiliser you apply if it is mineral Nitrogen. If the Nitrogen is in the form of the proteins in humus, it will not leach out, but if you cultivate the soil, it will tend to oxidise. As that happens, if the soil is moist, the Nitrogen will turn to nitrates and be easily lost into the groundwater. In very hot, dry weather, it may go off as ammonia into the atmosphere.

There may well be another difficulty. If the sort of plants you have are mainly woody and it is the woody prunings which you recycle, then they too will be short of Nitrogen.

You will gather from this that the worst thing you can do from a nutrient maintenance point of view, is cultivate. The alternative is to mulch and let the worms do the work.

Wastes which are Suitable, and Probably Safe
Repairing the loss is relatively simple; put nutrients and organic matter back into the soil by importing waste, preferably organic matter, from outside the garden. The most immediate place is your kitchen. The waste from there will provide almost all of what you might need including the Nitrogen. Whether you will have enough depends on how much nutrients you remove from the garden and how big your family is and how wasteful you are in the kitchen.

Kitchen Wastes
The micro-organisms in the soil and the compost heap are very much like ourselves in terms of nutrient needs, but they can, given time, digest almost anything. Below is a list of common kitchen wastes and their values.

Green Wastes including Fresh and Cooked Vegetable Waste
Good source of organic matter that will help to make humus. Green leaves especially valuable because they contain Nitrogen. Lawn mowings in the spring are the "dynamite" of composting with much Nitrogen and moisture. On their own, they will use up the Oxygen

quickly and "slump" down into a wet, soggy mass and likely to go anaerobic and smell. Mix mowings with other material and move it more often. In the autumn, remember that woody materials have much less Nitrogen in them.

Cut Flowers Past Their Best and Vegetable Waste
Ideal.

Uneaten Food After a Meal
Vegetable waste as above, meat good (contains much Nitrogen) but chop it up as it is slow to disintegrate. Make sure that the heap is tidy, preferably in a proper compost bin.

Paper and Cardboard
Not the best but by no means useless. Paper products are mainly Carbon which means that the micro-organisms can use it for energy and as the basic building block to make humus. However, to do that, it will be necessary to also feed some Nitrogen, in particular, and the full range of other plant foods including trace elements. These can come from the materials above in this list but don't over-do the paper. How much is "over-do"? Well, it is partly experience but the technology will give you the framework. (See "Carbon:Nitrogen Ratio in Chapter 5.)

Ash from Fires
Great if it is wood ash; that will give much potash and a good range of minerals (or "trace elements"). If it is coal ash, that will yield up the minerals but very little potash. All black coloured ashes have another effect (see next paragraph).

High Carbon Ashes, Charcoal
If the ash is very dark, really black, then it may have another value with a long term effect.

Studies of black soils (called t*erra preta*) in South America$_{34}$ have put another angle on Carbon and its value as amorphous or activated Carbon, usually (it is thought) derived from charcoal. This charcoal is sometimes now trendily called "Biochar". *terra preta* soils show a very high degree of sustainable fertility which, in areas of high temperature and violent fluctuations in soil moisture, might at first appear surprising. All of the technology here was developed within a national group of farms known as "Land Network" which is a consortium of farms which recycle wastes to land. Initial studies on these Land Network farms indicate that the Carbon in printing inks appears to be of similar value. The Carbon is capable of holding onto nutrients at a level similar to humus. However, humus is relatively easily and rapidly oxidised under conditions of high temperature and moisture but the pure Carbon is not. It may be that there is potential for the wider use of Carbon in soils in ways that we do not yet fully understand.

So, dark ashes are more likely to be beneficial than not.

Animal Manures
Great stuff with plenty of food for the micro-organisms and a fresh injection of its own micro-organisms (which may speed up the whole process). Horse manure is a useful source of potash. Poultry droppings are very rich in Nitrogen. Cattle manure is good in phosphate and Nitrogen. Yes, and the dog and cat can contribute, too. Smaller pets (rabbits, guinea pigs, hamsters – all of them) and all their soiled bedding can be included).

With all of these, because they potentially carry human pathogens, getting the temperature up is important, so do use a thermometer.

Values of manures vary from species to species, with what they are fed on and whether there is bedding with the dung and urine. Some guides are as follows and these are without straw or other bedding;

Manure	Likely % Moisture	Percentage in Dry Matter		
		N	P	K
Cattle	75 to 85	2	1	0.5
Pig	75 to 85	4	2	1
Chicken	60	6	6	3
Sheep	60 to 70	4	2	1

Animal Bedding

Cat litter may be volcanic ash which is inert and will add to the structure of clays. Sawdust is mainly organic Carbon (useful for humus production provided there is at least some droppings and urine). On this latter material, and as mentioned above, in old Victorian gardens, the old gardeners would encourage the apprentice gardeners to "pee" on the compost heap. Human urine (and of other animals, too) contains uric acid, which contains Nitrogen. Urine is (because of the acid) mildly antiseptic; so it is unlikely to carry disease if it is fresh urine (best not to leave it hanging about - get it into the compost heap).

"Fluff" from the Vacuum Bag/Cyclone

Great stuff, especially if it is from wool carpets (see below). Even if it is from synthetic fibres, these will be very valuable as fibres stabilise the physical structure of soils and allow them to breathe and manage water better.

And Some More Difficult Materials:
Cupressus Prunings
It is possible to compost these but they will process faster if they are shredded. All the coniferous trees, however, contain resins and these are very difficult for the micro-organisms to breakdown. Turn more often.

Wood and MDF
The wood is mainly Carbon and will need chopping or shredding into fairly small pieces. MDF is very interesting. There is, generally, a lot of misunderstanding surrounding MDF, chipboard and similar process wood sheet materials. Such materials are present in significant quantities in many of our homes, offices and the confined spaces in which we live. The most common binder or glue is ureaformaldehyde. Urea is used in large quantities, world-wide, as a Nitrogen fertiliser. The second part of the molecule, formaldehyde, if it were separate and on its own, is known to be toxic and potentially carcinogenic. The combined molecule is **not** toxic. There is an analogy with, say, common salt which is made up of Sodium and Chlorine – both of which are toxic if taken on their own. (Sodium is a silvery meal which dissolves in water and bursts into flame as it does so; if it were eaten, it would burn a hole though the body until it came out the other side. Chlorine gas is quite poisonous and a few whiffs will kill.) Yet we put the combined molecule on our food as "salt"; a flavour enhancer. Ureaformaldehyde is used both as an agricultural fertiliser and has been used for many years as a slow release Nitrogen source in many general potting composts sold in garden centres. The key phrase is "slow release" which has significance both

commercially in growing and environmentally in reducing pollution. So, if people tell you this material is dangerous, ask for their evidence and suggest that, if they were to be right, we all have much more risk from the furniture in our kitchens and offices. WHO (the World Health Organisation of the United Nations) has had a look at this and there is a clean bill of health. Everything has risks but, as far as we know at present, any risk from MDF is very, very low.

So, how can you use MDF in garden composting? Well, it is necessary to let the bugs get at it. One way is to shred it (which takes quite a lot of energy) or break it up a bit (say 15cm bits) so as to leave a broken edge. Relatively large pieces, say 150mm square can be buried under shrubs and perennials. That will give a slow release of Nitrogen. Alternatively, shredded material, or just broken up into 150mm squares, can be mixed with some farmyard manure or active compost made from this or other materials (these will add some micro-organisms and "seed" the process), then make sure it is quite damp and kept that way. A 15mm (half inch) thick sheet will keep on absorbing water until it is 50 to 100mm thick. Then it will break up and can be mixed in with the compost or direct to the ground. It is a really good source of Nitrogen – so, until you are used to it, not too much in one place and allow moisture and time.

Plasterboard
Plasterboard is a sandwich of Gypsum, which is Calcium sulphate, between two sheets of paper. Plants need both Calcium and Sulphur. The Calcium is especially useful in helping flocculate clays to give a better crumb structure, make them easier to work and plants will establish faster and grow better. However, the Calcium sulphate is fairly alkaline and so, use sparingly if adding

to a compost heap. Avoid plasterboard with an aluminium foil backing; that will not breakdown. The same remarks apply to waste plasters – most of which will be Calcium sulphate based. Better to put all of these direct to the soil where broken-up plasterboard will breakdown as it gets wet but it will look untidy in the meantime.

Old Mortar
Old mortar from an old wall will have been made from lime (probably "quick" lime which is Calcium oxide) and sand. We want the lime; it will help the compost, the soil and the plants. However, the mortar does need breaking up, preferably sieving, and remember how much it is diluted along the way onto the soil will affect the following plant growth. Calcicoles like it; calcifuges don't! Mortars are also alkaline and, therefore, go sparingly if adding to the compost heap.

Egg Shells
They won't breakdown in the compost heap, so break them up before they are put into the heap. However, there is Calcium in the eggshell and, as above, that is needed.

Coffee Grounds
Yes, the compost process and the soil will deal with this, too. Spread it out a bit or it will keep the micro-organisms up all night!

Teabags
Have you ever thought how tough a teabag is? You put it in boiling water and push it round violently with a spoon and even the couple of millimetres round the edge does not give in. How do they "glue" those edges to stand that? Why does the paper not disintegrate and let the tea leaves out? The answer is that this is high-tech

paper. There really is quite a lot of advanced technology required to design and manufacture a teabag. The base material is paper but the strength comes from the addition of a polymer plastic. Most gardeners will already know that teabags are slow to breakdown; this is why. Is it bad or toxic? Not a bit of it. The plastic is composed of just Carbon and Hydrogen. It will take time to breakdown but it will breakdown and do no harm. The fact that it is slow to disintegrate is a benefit because it adds fibres to the soil. (For fibre value, see Page 54.)

New and Used Carpet
Many gardeners have used carpet over many years to wrap round a compost heap to allow the process to breathe while letting rain in and reducing heat loss. These are all very important assets of carpet which is used as a carpet; just as a piece of material, still recognisable as carpet and used in pieces of a metre or two across.

All of that is a really useful application for carpet and it will eventually breakdown. But suppose we accelerated that breakdown by tearing it up. What is its potential value?

The majority of carpets sold in the last 20 years in the UK have been 80:20 wool:nylon. The backing has moved from jute, or hessian, to polypropylene. The immediate response is that it is the wool which is valuable. Wool is certainly of value because it is protein and that contains Nitrogen. However, there are other things of value. Look, for a moment, at the analysis below.

Typical Breakdown by % Weight of Average 80:20 Wool-Rich Carpets

80:20 with pp backing			% of total
facefibre	35%		
wool		80%	28%
nylon		20%	7%
Adhesive/filler	45%		
SBR		20%	9%
chalk		80%	36%
backings	20%		
pp		100%	20%

80:20 with pp backing			% of total
facefibre	35%		
wool		80%	28%
nylon		20%	7%
adhesive/filler	45%		
SBR		20%	9%
chalk		80%	36%
primary backing	5%		
pp		100%	5%
Secondary backing	15%		
jute		100%	15%

Notes; SBR is synthetic latex, and the PP (polypropylene) fibres are not UV stabilised.

Now, here is an interesting thing; the biggest single component is chalk! All plants use Calcium, the main value of chalk in soil. By the way, the adhesive will breakdown releasing energy to the micro-organisms. The nylon and polypropylene have real value and will be discussed in some detail below. So, carpet has a value in providing Nitrogen, Calcium and fibres.

Woollen Fabrics
Wool is pure Keratin – a range of proteins. Proteins contain Nitrogen in big molecules. Big molecules do not leach out, so this is slow release Nitrogen. So, old woollen sweaters? "Plant" one under a shrub or perennial. Alternatively put it in a compost heap; it will be easier to turn the compost if you cut the woollen fabric up a bit before adding it to the heap. It will be slow to breakdown but an active, warm heap will do it. It really does not matter if it does not breakdown completely.

The Value of Fibres in Soils
The last few paragraphs above have revolved round the value of fibres in soils. So, what is that value?

It is important to remember that all of the basic physical structure of soil is composed of mineral matter which is generally regarded as not biodegradable. Most soils, as any textbook, and the Environment Agency's publication "Understanding Rural Land Use" indicates, will also point out that the structure of good, stable, really fertile soils also depend on plant roots and "organic" matter to create a matrix which gives further characteristics to the mineral matter in soils in terms of stability gas exchange, moisture management, cohesion and erosion stability.

It can also be observed that many hydroponic crop production systems use vermiculite, polyurethane and many other synthetics as a growing medium to provide a matrix for root growth and a structure with a large surface area which the mycorrhiza can latch onto. These materials will normally be described as "not biodegradable". This is not strictly true in that all of these fibres, be they "natural" (such as lignin from plant roots) or "synthetic" (such as polypropylene from carpet backing) are all biodegradable but the process will, fortunately, take much time, maybe years.

Now, as all gardeners know, peat is used as a growing medium. "Organic matter" is regarded as providing stability and a range of growing qualities to soil. These materials contain cellulose, hemicellulose, holocellulose and lignin. All of these are big molecules, larger than the synthetics in the carpet mix. It is the fibres in peat which are the secret of what it imparts to soils it is mixed with or to a growing or potting compost.

Discussion on the Analysis of Wool
Wool composition is of "Keratin" proteins which, of course, contain Nitrogen and, in the case of keratins, useful amounts of Sulphur. In proteins of the Keratin group, several of the amino acids contain Sulphur and, of course, so do the amino acids which form some microbial proteins, plant proteins and animal/human proteins. These nutrients are clearly large molecules and therefore the nutrients are slow release and on crop demand.

Most people don't know this but, in most UK carpet composition, the most common material in the carpets is, as shown above, chalk which is used as a filler and stabiliser for the latex which glues the tuft to the Hessian or polypropylene backing. Again, the chalk contains Calcium which is a secondary, but very important, nutrient; second only to the major nutrients of NPK.

Jute used to be (and still is in what we will get for several years until currently new carpets get replaced) the most common backing and this is a fibre of plant origin, supplying Carbon and energy to soil microorganisms. Cellulose is, of course, mainly Carbon molecules in long chains of 5-Carbon rings. Woody stems, as in jute, will contain larger molecules of hemicelluloses with these long chains joined by cross linkages. There may also be some lignin. Lignins contain many hundreds of 5-Carbon ring, cross-linked chains. Lignin is closely associated in woody materials with cellulose and hemicellulose. The point here is that these "natural" fibres contain many hundreds of joined-up Carbon atoms. Most people accept that these fibres will take some time to breakdown in natural environments of soils of compost heaps, but these same people rarely question how long degradation takes.

The "synthetics" are large carbon molecules - actually with smaller chains than in many natural fibres. In the case of carpets, they are not "stabilised" against degradation by Ultraviolet light; so they are a little easier for the micro-organisms to digest. There is no question about breakdown and degradation; these do occur. The question is how long this takes and, with this material, in this application, at these dilution rates in the garden, it is a difficult thing to be precise about. Indeed, straw ploughed in may last several years. Whether slow breakdown is a good thing depends on circumstances but there are few or no known, likely or measurable deleterious effects.

The tufts contain, usually, 10% of fine nylon filament fibres. Nylon does have a very small amount of Nitrogen in its molecule. We do not yet have detailed knowledge based on long-term experience but the technology indicated that these fine filaments, when composted, and dispersed, will undergo significant degradation in maybe weeks rather than months in the soil. (Years are very unlikely.)

Polypropylene (Unstabilised). The table on Page 53 shows two analyses. The first is of jute-backed carpet – the majority of post-consumer carpet is currently like this. New carpets are like the second – for carpets which are "post consumer" (taken out of the house when replaced by new carpet) we expect a progression towards this material over some years. Ten or fifteen years ago, there was little polypropylene used in new carpet backing – it was all jute (or hessian). Now, the situation has reversed. As these new carpets age, it might be expected that the percentage of polypropylene in "waste" carpet will go up. What we have real caution about is how long polypropylene, if and when it is present, will take to breakdown under these circumstances. Even if shredded, it will take a long time, maybe years. This is not necessarily a bad thing, the fibres will act like the fibres in peat making the soil more stable, breathe better and hold moisture at the same time.

Why all this on carpets and fibres? Well, fibres are an important part of the structure of soils. Cultivation will accelerate oxidation of these large molecules and the soil will breathe less, water will flow through less, there will be less water held in the soil in a "healthy" way and there will consequently be more drought stress. Fibres make soils more productive and plants less subject to diseases. So use the materials round you to build fibres in your garden soil.

Municipal and Industrial Wastes
Just, for a moment, look farther than your garden fence and look at the national picture. Currently, Defra

appears to be obsessed with large scale plants which flies in the face of the Proximity Principle and is encouraging local authorities to "recycle" wastes through EfW plants; Energy from Waste$_{30}$. Whether you think that this is incineration by another name (which it is), and whether you think the energy gained means we save a small amount of burning fossilised fuels, is a discussion which appears to have been lost somewhere. What the route does do, without any doubt, is centralise waste collection, maximise trucks on the road and destroy organic matter which gardeners, nurserymen and farmers could grow flowers, food and biofuels with, so to save imports of mineral fertilisers and get trucks off the road by proximity principle recycling. So, anything you can do in the garden helps. However small, it helps.

Back in the early 1990s, the Enterprise Initiative of the DTI funded a series of studies looking at recycling urban wastes to land. Some of these studies ran costs into eight figures but there was one$_{21}$ which ran a total under £40,000 before the progress generated began to become self-funding - it resulted in the farmer-owner consortium Land Network which, between the individual Members, has, at some point, recycled over 100 materials, including many industrial wastes, even some liquid plastics (these can be spread out in the rest of the compost and are something similar to putting sugar on your porridge – just energy for the micro-organisms). It was done, is being done, successfully, safely and within the regulations in force at the time.

Part of that original study looked at how much "waste" there might be nationally which could be recycled to land sustainably. The figures were potentially unreliable but, after many discussions, including with what was then the Centre of Waste and Pollution Research at the University of Hull, the study concluded that there was possibly 100 million tonnes per annum. Land Network now concludes that the figure is higher, maybe much higher. The total land area of the UK is just over 24 million hectares but less than 20% is arable and just over 50% grassland and productive grass uses much Nitrogen fertiliser. Forestry would be more productive if compost were applied, too. So, say 10 million ha could be used for compost substitution for mineral fertilisers. At 25 tonnes per hectares of compost, bearing in mind that composting loses maybe a quarter of its weight, that means that the available land could use in the region of, say 30 tonnes per hectare of feedstock and a total of, possibly, 300 million tonnes. We have enough land nationally. If it can be done on a national, even international scale, safely, why send it on a truck somewhere else? That technology and experience can be added to centuries of gardeners' experience for every gardener to do it safely, productively and sustainably.

So, there is the big picture. How does your garden fit in? Well, there are a lot of gardens and if gardeners do it, and the farmers, and the foresters, and the national parks, and all land users who currently use mineral fertilisers, yes, we can change things. We can change the lives of the plants in the garden and the lives of our children.

Back to the definition of what makes a living organism alive and different from non-living; it is producing a waste. Where there are people, animals in captivity, businesses, factories, there will always be wastes. The concept of "Zero Waste" is only possible in systems where there is a closed loop which gives a sustainable cycle and that is very likely to involve going back to land. So, "the waste business" is centred round chimney pots; where there are people living and industrial activity. You, in your garden, may be a very small part of that but you *are* part and what you do does make a difference.

Municipal, Industrial, Business and Domestic Sources
There is an enormous range of materials which you can put into your soils. Generally, municipal authorities are a significant collection route for "wastes" and, as far as the European Union is concerned, Brussels has dictated progress by municipal authorities to collect and recycle. Brussels has further import statutory targets and failure by a local authority to meet targets will always result in financial penalty – sometimes tens of millions of Pounds. So what you do to recycle does directly affect the local taxes or rates you pay in every location in the UK and, indeed, in the EU. The main thrust of much of the regulation has been to direct separation at source, i.e. at the domestic producer household or local recycling sites. This "source separation" involves very significant expense and produces poor quality separation which limits remanufacture and delivers a level of safety which recycling to land has problems with. This problem of cost and less than satisfactory quality of output will result in developing alternatives which involve collection of whole, un-separated domestic waste. Source separation will fade out as these alternatives become available. However, source separation will remain longer than it should simply because of the intransigence and inflexibility of regulators and politicians.

Sewage has, historically, often been a municipal responsibility and, therefore, this source of waste has been identified, exploited and regulated. Now, putting raw sewage to land is illegal everywhere in the EU. What can be safely spread is sewage after treatment and this is usually called "biosolids" and that is a process product rather than a waste. It still has a stigma attached to it in the public and politicians' eyes for, perhaps, understandable reasons and, therefore, there are from time to time, moves to make it "disappear" by gasifying (pyrolysis) it or some other way of extracting energy by incinerating the Carbon in its organic matter. This, of course, is an enormous waste of its potential soil value as a very effective organic manure and a very safe one because of its ability to assist in reduced use of crop protection chemicals$_{35}$. There are a number of reasons for this.

Firstly, whatever form it comes in, provided it is from treated but whole sewage, biosolids have a wide range of nutrients, plenty of mineral trace elements and the good supply of Nitrogen is mainly slow release$_5$. So plants grow more steadily and strongly, with no flushes of rapid growth. Rapid growth is fleshy and more prone to aphid attack and, of course, the aphids are very likely to carry plant diseases. There is another most intriguing reason for an effect of biosolids on the health of a crop if the biosolids have not been heat treated or treated with lime.

Phages and Anti-Biotic Qualities of Biosolids
In the 1990s there was a desire, in what we previously called the USSR, to find the next generation of anti-biotics. The logic was glaringly simple. All species have something preying on them. There is the old adage of "every flea has a small flea on its back to bite them….." So, where are there lots of bad guys, because if we can find them, there must logically be good guys preying on them. The obvious place to look, for the researchers at

the time, was in sewage. They had plenty of that and it was already there with a "negative cost" i.e. they needed to get rid of it. Sure enough, when they looked, they found human pathogens. When they looked harder, they found the good guys attacking the pathogens. The good guys, anti-biotics, were identified as phages. Phages are just sub-optical for the naked eye but can be easily seen with an optical microscope; they look like multi-legged spiders. This, then, is one of the reasons why untreated sewage products can be beneficial in reducing crop disease. Sewage does need sensible handling, especially if there are children about. Those who handle it at all may be well advised to consult a doctor and have a series of vaccinations as a precautionary protection. However, some garden centres sell tubs of dried biosolids and they have everything except (because of treatment) the pathogens and the phages. The important thing to note is that, for gardeners, all dung and urine has value for the nutrients it contains, not just the NPK, but the full range of elements including the traces necessary for healthy life. They are almost always somewhere near the right balance, or at least not dangerous simply because they have come from a living organism. These excreta also are likely to have a lively microbial population which is a two-edged sword demanding care as well as giving advantage.

Municipal and Industrial Recycling
Most businesses and industries are significantly behind municipal authorities in the restrictions and pressures imposed on local authorities by the regulators. Nevertheless, because of rising landfill costs, all businesses will become increasingly active in managing their wastes better. It makes sense that raw material should be recovered to reuse in profitable production if

possible. The key word is "profitable" simply because handling anything, processing anything, doing anything, involves resources and the cash to pay for those resources has to come into the equation somewhere. That source of cash depends on the general business environment of prices of raw materials, regulation and related taxation. It is also relevant to think medium and long-term in that the climate of what is "acceptable waste" is changing. Whatever anyone thinks about the word "profit", it is as well to remember one principle; if a situation is environmentally not sustainable but is financially sustainable, then it can go on for hundreds of years, but if the reverse is the case with a situation being environmentally sustainable but not financially sustainable, then it dies today. The fact is that true sustainability involves both environmental and financial balance.

Recycling industrial wastes to land is a real technical option limited mainly by regulation. Now is the time to look again at recycling the what, when, where, who and how - the why is a rhetorical question. Everyone in the waste business knows that it is possible to recycle green wastes (from gardens) to land. However, there is a limited supply of that and industry is facing dramatic rises in gate fees and restrictions in going to landfill or to high-tech processing. The plain truth is that the most high-tech processing operation yet designed by man is nothing, nothing to match the complexity, thoroughness and safety of a compost heap and a fertile soil. So, whatever the Local Authority does, whatever technology is developed to do clever things in a big shed on an industrial estate, it is still infinitely preferable to keep trucks off the road and recycle as much as possible for every household to that household's own garden soils.

The micro-organisms in a compost heap are mainly bacteria, Actinomycetes (they give wet woodland its attractive smell) and fungi. The soil is much the same but usually with a shift to the fungi. The numbers are astronomical; a researcher in the USA did an estimate: "an acre of arable land has a weight of micro-organisms equal to the weight of an adult cow". The numbers in a cubic metre of active compost runs into trillions and the diversity is mind-boggling and highly variable (depending on what material is there and the stage of the breakdown). Incidentally, the predominant group of fungi in composting is of "bread moulds" or Penicillins! There is another advantage of recycling; British farmers import around £1 billion worth of mineral fertilisers pa.

Feed the Bugs
The micro-organisms in compost and land are different from us, of course. However, in fact, they are remarkably similar in terms of dietary needs. They are different in that they can tackle much bigger molecules than we can[32]. Just as we might put sugar (a 6-Carbon molecule) on our cereals for breakfast, the micro-organisms can tackle the lignin in wood (with an almost unlimited number of Carbon atoms in the same molecule) and they never stop eating - 24/7.

As an example of this capability, it is interesting to look at one of Land Network's farms[21]. (The group is a consortium of farms which recycle wastes to their own land.) The example farm takes wood from local municipal and landscape gardening sources. Cellulose is made up of long chains of maybe 3000 monomers. (Monomers are groups of 5 Carbon atoms in a ring.) So there are around 15,000 Carbon atoms in one of these chains. Lignin is made up of cellulose chains lying next to each other and joined by cross-linkages. This means that xylem (the "wood" behind the bark) in a tree may be almost just one molecule - everything is joined together. The number of Carbon atoms joined together is, obviously, enormous and almost so big that it is beyond ordinary number-comprehension. This farm also takes in PVA (Polyvinyl alcohol) as a hot/warm liquid (it solidifies on cooling). As a liquid, it is comparatively easy to disperse it through a large mass of compost (preferably active and, therefore also hot). This is a big Carbon chain alcohol with many thousands of Carbon atoms in each chain. Even so, lignin, the basic molecule of wood, is even bigger with many cross linked chains. As soil and composting micro-organisms can crack lignin, of course they can crack PVA; it is just like humans putting golden syrup on porridge. The BOD (Biological Oxygen Demand) - see below, of PVA is enormous, so it is necessary to keep turning the compost but, as a result, the bio-activity rises rapidly and the process is faster. It helps make very good compost. PVA is a plastic; so we can and do compost plastic provided the format allows the micro-organisms to attack it.

A note on BOD which is a measure of how polluting a material is; BOD is a measure of how much Oxygen will be needed by the micro-organisms to break it down - they breathe just like us to turn the big Carbon molecules into Carbon dioxide and water. We do it with the sugars and fats in our bodies; they can do it with the big molecules in plastics if they can get at them. They can't tackle plastic sheet, except very, very slowly because the surface area is very low by their standards. It is different if you can spread the plastic molecules out, as this farmer does with the liquid PVA spread through the rest of the compost.

So, compost heaps$_{27}$ will crack almost any organic molecule. ("Organic" to a chemist means the molecules are based on Carbon chains.) Given a balance of other necessary food molecules, they will crack 100% of the Carbon chains in the feedstock in time. Does it matter if there is any residue before spreading to land? not really, provided that the soil is biologically active. (That means that it already has a reasonable amount of organic matter in it.)

To say that the Carbon chains will be cracked is not the complete story. The micro-organisms make new Carbon chains; as hydrocarbons, carbohydrates and proteins. When the micro-organisms die, these molecules form a dirty black tarry substance - initially very similar to crude oil. (Give it a few million years with no Oxygen and some pressure and it will actually be crude oil!) This dirty black stuff (some soil scientists do call it DBS) is actually what most of us call "humus"; it is the material which gives soil its black colour.

Generic Values of Materials and Values to Land and Crops
Within the context of this book, then, there are some generalities which give some guidance. If the waste has much Carbon in it, then it can be composted or burned (for energy recovery). However, if the Carbon is present in large chains in the molecules, it is valuable as an energy source for the micro-organisms. If the waste has any of the useful garden plant nutrients, then it can be composted and recycled to garden soil. These nutrients are the major ones of Nitrogen, phosphates and Potassium salts, followed closely by Calcium, and Sulphur, then Magnesium and a whole range of "trace" elements which include most of the "heavy metals" (including Copper, Zinc, Manganese, Cobalt and many others) of which regulators generally shy away from. The

truth is that evolution has created an ecological balance which not only can tolerate most things, they actually need most things. For example, as you read this, your blood will contain around 10ppm (parts per million) of Arsenic. If it did not, you would probably feel sick and might well be dead. However, if your blood contained 1000ppm, you might feel sick but you'd probably be beyond that and be dead. The suitability for recycling to garden soils, or indeed for energy crops, forestry, or land reclamation depends on how far the material will be dispersed and on detailed knowledge of soil science. This area of technology of environmental management is known as "Dispersion Technology".

There is one other possible way of judging safety; if you could eat it, possibly/probably, so could a farm animal, so could the micro-organisms in a compost heap. Humans have difficulty in digesting the cellulose in garden wastes but ruminants do not. There is a remarkable similarity[26] between the micro-organism universe of a compost heap and that of the rumen of cattle, sheep, goats, antelope or any other ruminant (see below).

Logic says that the high cost of source-separation (such as households putting out a "green bin" for collection by the local authority), so favoured by the regulators, will eventually be superseded by more dependable recycling routes. There are two problems with source separation; the high cost and the fact that it cannot be trusted. If, for example, a member of the public puts a small NiCad battery in a bin and it ends up part of a 10 tonne load of "green waste" destined for a large composting operation, there is no way anyone is, in practice, going to find that. What happens in practice is that the resultant compost is spread out far enough to reduce the risk of pollution from that battery to an acceptable level. That can also be achieved without the cost of source separation and the technology for central separation in a factory (often called a MRF - Materials Recycling Facility) is there, ready and waiting.

Specific Materials and Value to Garden Soils
Generally speaking, everyone thinks that "green waste" from domestic gardens makes ideal compost. It might but might not. Garden waste from high density housing built maybe within the last 5 years and the waste collected in, say May and June, will be high in Nitrogen and likely to be of high value in composting and to the land onto which it will be eventually spread. However, garden waste collected in, say, October and November, from detached houses built 15 to 20 years ago, is likely to contain much Cupressus and, if there is much trunk timber in the sample, there will be much energy used in shredding it and not much Nitrogen in the resultant compost product. There is also likely to be a problem in the slow breakdown of the resin in the trunk timber from conifers. If this material is not composted with both another Nitrogen source and significant moisture added, then it is quite likely that crops "fertilised" with the resultant compost will be negatively affected$_{22}$.

Moisture in Compost Processes
All composting operations are driven by micro-organisms which, of course, are alive. All life as we know it is dependent of moisture. Without moisture in a compost heap, it will end up as with many farm and domestic processes, i.e. as preservation rather than processing. Composting garden waste without enough moisture will produce a "woody hay". It may be dark coloured and friable but the process of pathogen kill, of immobilising

the soluble nutrients, of killing weed seeds, will all be less than complete.

Having recognized that, there is a practical difficulty in deciding what level of moisture to target to get the best process. Firstly, materials delivered to a compost heap may have too little or too much but they have what they have; adding moisture may be easy enough in principle if it is available (in hot climates, it may not be), and removing moisture could be difficult (such as being so wet that the compost activity has difficulty in starting and generating the temperature that will drive off the excess moisture). Partly, it is a question of "feel". Health and safety demands careful examination to avoid splinters and cuts; common sense must prevail. However, skin touch gives a guide. Ideal is 50 to 60% moisture. It will feel neither dry nor wet to the touch. Too dry will fall apart easily, too wet will cling together more and feel wet.

There are two important functions which will dictate where the optimum is; aeration and adequate moisture for process. Firstly, if the input material for composting is well structured, such as with coarsely shredded and woody garden waste, then there will be enough air for the process to start and to build up heat. Further, the windrow or heap will be able to breathe, much as submarine snorkels, letting heat out and new air in. From a garden point of view, a range of particles up to 100mm will give a significant amount of flexibility in how often the heap needs turning. If there are large particles, then the heap will, in a wide range of circumstances, breathe on its own and you may go off onto other pressing tasks. As particle size decreases, then that flexibility is lost but the process will progress at a faster rate if the heap is turned to let the air in. Also, where there is a significant proportion of large particles, the material could be very wet and the process still progress satisfactorily. If, however, the material is mainly something like grass lawn mowings or white onion flesh, then it will slump very easily and run out of air very quickly and go anaerobic and smell offensively.

So, where possible, it may be a mixture of materials which gives optimum process control. Commonly, this will have moisture contents in the 40 to 60% range. If the material has good air spaces connected to external air, that could be as high as 70% but especially with poor air spaces, above this there will be an increase in odour risk. As the moisture gets below 40% there is likely to be process slow down and dry out.

There is one area where moisture becomes in significantly greater demand and that is with some industrially-prepared materials such as MDF. Fibre boards can absorb several times their own weight of water before expansion takes place which disintegrates the board and allows the composting process to proceed. By the way, good peat will absorb up to 16 times its own weight in water (which is one of the major reasons why we use it as a potting material). Composts made in the garden from wastes will approach that but rarely be quite as good; they will usually hold in the range of 5 to 10 times their own weight.

This question of moisture in the compost process will be referred to again, with some test figures, in Chapter 7.

Chapter 5
How to Manage the Nutrients in Waste
Nutrients in Waste. C:N Ratio, Trace Elements and Nutrient Balance.

Nutrients in Waste
All materials, not just wastes, contain, at least to some extent, "food" for micro-organisms. Natural ecosystems are insidiously capable of reconstructing almost everything through closed loop processing. Even the Titanic, sunk 2 miles down, is not rusting in the absence of enough Oxygen, but it is being eaten by bacteria. It may take several hundred years, but they will do it. This demonstrates something of regulatory significance. Very often, those not skilled in the art, with limited technical knowledge and/or experience, will say something won't work or is impossible. Most things work in nature; it will handle anything given time. The question is more about how long it will take.

There is a word of caution in thinking about Carbon and its use in soils. Carbon certainly can be involved in large molecules which are, in the language of an organic chemist, "organic". Carbon may also have a very significant effect on the fertility of soils when it is present as pure Carbon, in its amorphous or activated state, as shown in the *terra preta* soils of the South Americas.

Micro-organisms certainly need food and they need moisture and gas exchange - if we are interested (as we are) in aerobic composting, they need Oxygen from the air and need to be able to breath out Carbon dioxide, just as other forms of life, including humans. But what do we mean by "food" and how flexible are the micro-organisms? Bearing in mind that farmers have been ploughing in 6 to 8 million tonnes of straw per annum for over 20 years in the UK, does a text-book view of "Carbon:Nitrogen Ratio" really matter?

The "food" that micro-organisms need is basically Carbon (for energy - just like sugar and carbohydrates are to us) plus all the other nutrients that our own bodies live on. The Carbon can be in the form of almost any organic molecule (i.e. any molecule containing Carbon linked to other Carbon atoms, Hydrogen and possibly many other elements). They also need a balance of Potassium, phosphate, Calcium, Sulphur, Magnesium and the full range of trace elements, just as we do. So what's different? Answer, not much except the ability to use a wide range of Nitrogen sources to build their own body proteins but, unlike us, they can use non-organic forms including, in the case of the bacteria attached to the roots of legumes, atmospheric Nitrogen. They also have the flexibility to be able to adjust to wide variations in the concentration of organic Carbon molecules.

In discussions about C:N ratio, a range is often given, commonly 20 - 30 to 1, without reference to whether this refers to the start of the compost process or where the garden wants to end up with in the finished compost. There are many text-books, academic papers and standards relating the proportion of Carbon to Nitrogen in a compost process and product.

Basically, micro-organisms need much Carbon (for energy) and not so much Nitrogen (to make cell body proteins). Many "authoritative" guides will give, as a guide, an ideal of 25 (of Carbon by weight) to 1 of Nitrogen. Alternatively, a range of 20 - 30 to 1 is usually a better guide to what actually works. The truth is that it can be much wider than even that and this is because the micro-organisms will use up some Carbon in the process. Further, the more difficult the process, then the longer it will take and more energy, and therefore Carbon, they will use up.

As a guide, the following gives an indication of the C:N ratio of a number of materials which may be "on offer" to a composting operation;

Material	C:N Ratio
The best top soil	10:1
Humus	10:1
Kitchen food waste	15:1
Vegetable wastes (factory)	20:1
Lawn mowings (UK May)	10:1
Lawn mowings (UK August)	25:1
Tree leaves (autumn brown)	35:1
Woody prunings	400 to 600:1
Manure (stable, high straw)	50 to 70:1
Manure (cattle, low straw)	20:1
Wheat straw	70 to 150:1
Oat straw	50 to 100:1
Sawdust	150 to 800:1
Newspapers	150 to 900:1
Pig manure	10 to 15:1
Farm yard manure with straw	15 to 25:1
Horse manure without straw	25 to 50:1
Hen manure	5 to 25:1
Seaweed	20:1

As an example of handling high Carbon inputs successfully, in the mid 1980's British farmers faced a ban on burning cereal straw behind the combine harvester. The question was; what would happen when around 6.5 million tonnes of straw was ploughed in every year? Put another way; how could this be best managed?[20]. By the mid 1990's, every farmer was doing it without a second thought and yet, the straw had very little Nitrogen in it and the C:N ratio was well outside the figures discussed above; straw will, depending on variety and season, have a C:N ratio into the 70 to 100 to 1 range.

Generally speaking, decomposition by micro-organisms is much the same process as digestion by ourselves in that it takes energy. As we metabolise sugar during exercise, they will metabolise Carbon-based molecules to produce the energy to drive the process and produce Carbon dioxide. They need Nitrogen to build cell proteins but they need much more Carbon than Nitrogen. In processing the Nitrogen, they will metabolise Carbon and the more difficult the digestion, the more Carbon they will use up. As the C:N ratio gets much above 30:1, then the process gets more difficult and uses up more Carbon and takes longer. There is one more factor to complicate the issue and the judgement on how well the process will go and whether more Nitrogen will be needed. With composting hedge cuttings, for example, some of the Carbon is in the form of cellulose and will quickly breakdown if there is enough Nitrogen in the system (such as if there is much green leaf and not much woody stems). Further, there will be much Carbon present as more complicated molecules with more stable cross linkages in the Carbon chains, such as hemicelluloses and lignins (such as if there is less green leaf and more woody stems). These will take longer to process and may require more Nitrogen for the micro-organisms in the short run.

Chapter 6
The Structure of Soils
In the Surface Soil; Sands, Clays, Loams, Chalk, Organic Matter, Fibres; The Soil Profile and Root Growth.

In the Surface
The word "texture" means different things to different people. For most, it refers to the size of the soil mineral particles; sand, silt and clay. A reasonable mixture of these is often referred to as "loam".

Internationally, there are a number of definitions of what each of these groups are in terms of particle size. As a guide, the following is a British garden guide which is not far from the international standards;

Gravel	above 2mm
Coarse sand	from 2mm down to 0.6mm
Fine sand	from 0.6mm down to 0.2mm
Silt	from 0.2mm down to 0.002mm
Clay	from 0.002mm downwards

If you have what might be called "clay" soil, it will not be all clay within the above definition. For example, an Oxford Clay will be in the region of 15% sand, 30% silt and 45 to 50% clay. (There will probably be some other things to make up to 100% such as gravel, stones.) If you have a sandy soil, then the proportions of each component will shift towards the sand but not, in nearly all situations, completely to 100% sand. Similarly, it is rare for a "clay soil" to be all clay in the text-book, engineering sense. A "loam" is a mixture of all of them in more or less equal amounts.

Sands feel "gritty" to the touch and do not hold onto nutrients when subject to rain or irrigation. Sands do not hold onto water either and plants will suffer drought stress more easily. Clays feel smooth to the touch, smear easily and do hold onto nutrients much better. Clays will hold onto water (so drought stress is less likely) but these soils will also hold onto too much moisture and be difficult to work in wet weather and slow to warm up in spring. All soils with low organic matter levels will compact easily, reduce root growth, be difficult to work, show higher levels of plant disease and generally garden less well$_{28}$.

Historically, when a soil was substantially clay or sand, there was a widely, but laborious, practice of "marling". This involves adding sand to a clay, or clay to a sand. The aim is to produce the mixture we call loam.

There is a much less laborious and more sustainable way of improving the long term fertility of soils; adding organic matter. So, an alternative in farming used to be "green manuring" where a crop of, say, vetch was grown over the winter and then ploughed-in in the spring. In the garden, the compost heap will significantly change all the negatives of relatively pure sands, or clays and will, even in fertile loams, add all the positives we look for, and regular, preferably heavy, dressings of compost will result in significantly less crop disease.

There is a remarkable feature of clay where organic matter is added. Clay and humus can combine to form very stable, complex substances which appear to have greater ability to withstand breakdown by micro-organisms and oxidation by cultivation$_1$.

Chalk
Anyone who gardens on chalk will tell you that chalk, and indeed any high lime-content soil, can lock or

immobilise other important trace elements. The problem is that any particular soil has a specific capacity to hold ions (those charged particles again). Suppose, for the sake of argument[20 and 25], a soil could hold 100 of these negatively charged ions. These could be ammonium or any of the metals. Now, a normal fertile soil might hold, say, 65 out of the 100 spaces for these ions, occupied with Calcium. Suppose you have a chalk soil or put too much lime on other soils, and the Calcium now occupies, 75 places. That soil will now become heavier and more difficult to work but something else happens. Before the extra lime was added, there were 65 places occupied by the Calcium ions and that means there were 35 occupied by something else. They would be Potassium and Magnesium and then all sorts of other ions including Sodium, Cobalt, Manganese, Copper, Zinc and more. Put the lime on and the available spaces drop from 35 to 25. Where do the rest go? They either go into an unavailable form, a precipitate which may become unavailable to plants (but they are still there), or they get flushed out by rain into the groundwater and are lost to the surface soil and your plants and crops. Chalk soils are like that and, maybe, more so.

Undisturbed chalk soils generally drain well and can hold some water. However, work them in the wet and they may become very sticky and difficult to cultivate and afterwards become very difficult for root growth and cap easily.

Adding organic matter will dramatically change these characteristics, mainly for the better, but never totally eliminate the problems. Organic matter will certainly increase the capacity to hold nutrients and allow more to be available to the plants and crops. In the wet, high organic and high Calcium may make these soils very sticky and they are best left alone.

Fibres
Fibres, whether they be from peat, plant roots, compost or synthetic sources, will add structure to a soil and help it drain better and breathe better. Put scientifically; they will allow gas and water exchange more easily but slow down run-off of both water and nutrients; it will be a more controlled movement with less extremes. There is much less likely to be a problem with flash flooding in the UK if there were more fibres in the soil. The UK hills up to a height of over 1000m were covered in trees until around 5000 years ago. Over-grazing has allowed much of our upland soils to be eroded and carried by rivers into the sea, lost to the land forever[29]. Adding fibres to soils will significantly reduce that erosion risk. One of the advantages of the "fluff" from the vacuum cleaner bag is that, with modern carpets, more of the fibres are nylon and polypropylene which do breakdown but more slowly than some natural fibres. Greater fibre stability and longer life is a significant advantage especially if the soil is cultivated regularly and thoroughly.

Organic Matter
So, organic matter in general gives three things to surface soils; better physical characteristics, better biological activity with less crop disease, and when the organic matter is humus better chemical characteristics (with less loss of nutrients to groundwater[24]).

The Soil Profile and Root Growth
With the exception of a regularly short-cut lawn, most garden plants will put roots down 1 to 2 metres in one season[1]. Provided, that is, that the soil is structured to allow that root growth. It is quite possible to grow many

plants, bedding plants and quick-crop vegetables for example, in a shallow soil, even as little as 150mm, provided it is reasonably well managed, even in a sand with adequate nutrients and moisture. It will do better with good organic matter. It will do better still if there is greater depth. If there is to be less labour and less management time put in, then that organic matter and more depth become more important because the roots can go deeper to tap greater reserves of nutrients and moisture. So soil depth will save work.

Some soils will produce a "pan" or naturally dense layer. Some of these soils, known as "podsols" are predictable in certain acid soils. Some of these pans are man made; some by gardening but, very commonly in a garden near a building, by the original builders. It is general practice on a building site to wait until the building is done and then bury everything on site under a thin layer of soil; and put turf on top and call it "garden". Whatever the reason for it, the garden will do better if any such barrier is broken. That means digging.

Digging is most obviously done with spade or fork, preferably stainless steel and sharp (it's easier). However, there can be a surprising amount done by worms and roots – both of which will be encouraged by adding organic matter to the soil. Here is another thing which organic matter helps with and that is the reduction of barriers to root growth further down the soil profile. It is not as quick as mechanical digging but it is much easier and it does last longer.

Chapter 7
Liquids and Compost Moisture

It is important to remember that the compost process needs the micro-organisms, the food, Oxygen and *moisture*. Green waste on its own, over a year's supply of materials including relatively woody inputs, will certainly go black and friable in most circumstances but the process is more likely than not to be incomplete with much of the originally soluble nutrients still soluble. It is only by thorough composting, which needs moisture to push it all the way through, that the solubles will turn into humus and reduce the leachable nutrients to near zero.

How many times have you, in your lifetime, seen a hosepipe ban inflicted on gardeners? Gardens do need an enormous amount of water and hosepipes do put a significant extra demand on our reservoirs. Did you know that parts of Suffolk have less than 375mm of rainfall a year and that is the United Nations' definition of desert? While water may not be a problem for many in the Western parts of the UK, in most years, it is an increasingly expensive asset which will probably (maybe eventually necessarily and permanently) be restricted for garden use[19]. The reason for that is that the building of new reservoirs has not kept up with population growth and industrial use.

Compost Can Help your Land.
People in arid climates outside the UK may find it difficult to believe but demand for agricultural irrigation water in the UK can at its peak be up to 25% of national demand and 50% of total demand in the Eastern region of the UK. It is probably similar in most countries with developed and productive agriculture where water use

for irrigation and stock drinking is a significant proportion of total demand overall. Like all water use, it will be under continued scrutiny as demand outstrips supply. The current over-emphasis on Demand Management avoids the fundamental issue of reservoir building. Reservoirs have a cost and generate political problems from the NIMBY (Not In My Back Yard) and Green pressure groups.

There is an alternative which will help, but not cure, the problems involved with shortfall of conventional reservoir capacity. It is to do with better management of the land itself. Waste containing large Carbon molecules can act as a large, well-structured sponge. Composts, therefore, can have a major effect on successful soil water management. Composts can hold up to ten times their own weight of water and, therefore, addition of large tonnages can be used to form "topsoil reservoirs"[19].

Soil as a Reservoir
According to Defra statistics, there are around 10 million total agricultural hectares in the UK and 5 million arable. Research in the early 1990's by Land Network International (LNI) in East Anglia[21] studied a contractor, Alwyn Moss, who short-composted shredded newsprint used as bedding in the stables at Newmarket. He spread this compost at 250 tonnes to the hectare (100 tonnes per acre) onto a blowing sand at Mildenhall and grew fodder beet *without* irrigation. The area normally has only 380mm of rainfall per annum and neighbours needed 150mm of irrigation (1500 cu m per ha) to grow the same crop. This under-composted material was tested and shown to absorb 5 to 10 times its own weight of water during the autumn and winter rains and give it up to the crop in the following growing season. As an

example of the scope, 1 million hectares of treated arable land delivering this sort of holding capacity would give a soil reservoir figure of 1.5 billion cu m. Building such a soil reservoir would take, on these field research figures, 250 million tonnes of composted material or, coincidentally, a compost yield from something less than the total not-bio-unfriendly, high volume/low value wastes currently collected in the UK in the form of Municipal Solid waste (MSW or garbage), industrial and business wastes. The corresponding figures for a single small garden are, of course, comparatively tiny but they are no less significant; what we do in individual gardens matters.

Nutrient Run-Off Reduction
First of all, the problem. On the Environment Agency's figures, maybe 40 to 45% of mineral Nitrogen fertiliser goes straight into groundwater when it rains. It may be more on a sandy soil and may be less on a clay. Whatever the soil, whatever is spent on Nitrogen fertiliser in a year, approaching half of it is lost by rain to that garden or farm. It costs someone else more than that to get it out again. To that extent, the EU is quite justified in calling for a re-examination of what we are doing in the UK. Unnecessary nitrates in our rivers and water supplies for human consumption are certainly an expense and a potential risk to human health.

Where there is a problem is in that regulation by Defra and the Environment Agency to date treat all Nitrogen as the same. It isn't. Their biggest error is to treat all organic Nitrogen as the same. It isn't.

If ammonium nitrate is added to a sandy soil, then neither the nitrate nor the Ammonium will stay there very long in rain. On a clay soil, however, the colloidal properties of the clay will hang onto the ammonium part. So, put ammonium sulphate onto a clay soil and you will keep the Nitrogen in the Ammonium part for longer and lose less because the plant will get it, given time. Put ammonium sulphate on a high organic soil and you will be even better off because the humus in soil will hang onto both the nitrate and the Ammonium parts. Put the Nitrogen on as organic Nitrogen and you will lose very little. Put organic matter into a compost heap and compost it thoroughly and you will lose almost none in rain, not ever, even if it is a 1000 years before any plant needs it.

Table 7.1 is a record of composts from a small group of Land Network farms. The composts were tested in the Environment Agency's own laboratories and subject to a wide range of tests – the report here is limited to total Nitrogen and nitrate from a leachate test$_{41}$.

Table 7.1
LAND NETWORK
NITROGEN / NITRATE RESULTS ON COMPOST TESTS

Date	Site	Total N mg/kg	Leachate mg/l
26/06/2006	Crows Nest Farm	13400	<1.00
31/07/2006	South Elkington	16600	14.7
04/08/2006	Beech Tree Farm	9600	<1.00
22/08/2006	South Elkington	1280	1.04
06/09/2006	Faldo Farm	18700	1.62
21/11/2006	Beadlam Grange	1300	25.8
24/04/2007	South Elkington	14000	16
16/07/2007	Beech Tree Farm	14000	<2.2

The total Nitrogen in the samples varies, according to the original feedstocks and the individual farm process. The highest values are not insignificant; these are quite good fertilisers. The leachate tests also vary but some get below 1mg/l or parts per million. These low readings show, conclusively, that the process has eliminated nitrate pollution of the groundwater. This claim of "eliminated" is justified not by showing the reduction was to zero but by showing that the level was reduced to a level which was not just ecologically tolerable but necessary. This is so because all ecosystems depend on a low level of leakage – if there were zero leakage, then the weeds in the river would not grow and the fish would die of starvation.

There was, however, a remaining question of why were some of the nitrate figures *not* down to around 1ppm. Investigation showed that those compost operations were short of moisture. As referred to in several places elsewhere in this book, moisture is necessary for the micro-organisms to work and complete their consumption of potentially soluble nutrients. Excess moisture may have a point risk if the compost operation is on bare ground without concrete or asphalt under it. However, as this small piece of research shows, adequate moisture is fundamentally necessary for the micro-organisms to complete their work and protect the groundwater from pollution and the gardener from economic loss. It is of some interest that gardeners may choose, if their soils are not fully supplied with nutrients from the available compost, to add supplementary Nitrogen and other potentially soluble nutrients to the compost heap before the process is complete. If mineral fertilisers are added to the compost heap, not later than the middle of the process, then those soluble nutrients will be eaten by the micro-organisms and built into the humus. This way, the process will eliminate economic loss of nutrient by leaching by rain or irrigation. At today's fertiliser costs, this is potentially quite important.

What is the ideal moisture content? Well, yet again, it will vary depending on what the feedstock is, what the particle size is and how long the process is allowed to run. Generally, it is likely to be best in the region of 40 to 60%. More than this and it is likely to go anaerobic, slow down and probably give off more offensive odour. Below this it is likely to give an incomplete process. Generally, larger particles will allow higher moisture content without going anaerobic. Woody materials will usually allow higher moisture content but small particles of high Nitrogen-content materials, such as grass cuttings, will go anaerobic at much lower total moisture content simply because of smaller inter-particle spaces for Oxygen which will be used up faster by the readily available Nitrogen.

It is important to note that the heat generated by the process will use up and drive off very large quantities of water and liquids may need to be added ***during*** the process (not all at the beginning – spread it out) in order to avoid going anaerobic at any stage but still ensure completion of process.

Chapter 8
Compost - The Finished Product
Fertiliser, Soil Improver, Mulch

Summary on "Humus"
Put in scientific terms, humus is part made up of hydrocarbons, carbohydrates and proteins. These are large molecules which are not soluble in rain. Also, humus has three or four times the colloidal capacity of clay and will hang onto not only cations (which are the electrically charged particles of Calcium, metals such as Potassium and Magnesium, and also ammonium) but also anions (the particles of nitrates, sulphates and chlorides or "Muriate"). Organic matter is turned into humus by bacteria and fungi during composting or incorporation into the soil as, for example, when trash is dug in. However, incorporation results in a slow processing into humus and some nutrients may be leached out. The compost heap is a good "buffer" to get everything changed to humus before application to land.

So, what about Fertilisers, Soil Improvers and Mulches?
It is quite common, especially for the regulators, to ask if the compost product is a fertiliser, a soil improver, or a mulch. When it is dug in, it is a fertiliser and a soil conditioner. If it is left on the top, it's a mulch but the worms will take it down and then it is incorporated and so it is a fertiliser and a soil conditioner.

Now, "soil conditioner" is a somewhat abused term and might mean almost anything. However, there is a very important interpretation of what it means. Soils in "good condition" have a number of important characteristics; physical, chemical and biological.

Soil Physics

Soils have a number of physical functions. Firstly, and perhaps most obviously, anchorage to stop plants blowing away and, less obviously, stop itself blowing, too. The sandy soils of Suffolk in the UK will, if not covered by a crop and if they have dried, blow quite easily. Parts of Suffolk have less than 375mm rainfall and less than this is the United Nations' definition of "desert". Yes, the soils of Suffolk will blow and it will darken the sun and strip the paint off parts of a car. Organic matter is one of the cures and that organic matter is better if it is large particles, not too well rotted and plenty of fibrous material. Fibres are an important part of soil to hold it together and stop it compacting down, especially when wet, and squeezing out the air. This applies to all soils, not just the sands. In clays, this ability to hold the clay open to allow gas exchange and water movement is vital for crop growth. Composts with large particle size, not completely rotted, will do this and large amounts of compost will be better than less. In fact, there is no environmental limit to how much compost can be put on. If the Fens were originally up to 10 to 15m deep and those reserves never polluted the groundwater, you adding a bit to your garden is safe, too.

If there is a good physical structure, then the chemistry and biology can work too.

Soil Chemistry

Pure sand is silica and near inert. It will hold little or no nutrients but, if not compacted, most sands will allow relatively easy root penetration. At the other extreme there is the sort of clay that can be used to make tiles; it is heavy, very difficult to achieve reasonable gas exchange and moisture movement and roots will have difficulty because of that. However, it will hold nutrients. Clay is made up of very small platelets which are negatively charged and, therefore, will hold onto ammonium particles (called "ions") and other metal ions such as Calcium and Magnesium. These particles are ions and of a particular sort because they are positively charged and, therefore, called "cations". Clays are not very good at holding "anions" (which are the negatively charged particles) such as phosphates and nitrates. That is why sands will hold very little nutrients and clays much more.

Organic matter, when it is turned into "humus" will hold not only four or five times the nutrients that clay will, it will hold both the cations *and* anions - all the metal ions and the nitrates and phosphates. Humus is staggeringly better all round. It is possible to work without humus, but it takes more nutrients, more muscle to dig the soil, and more crop protection chemicals, too. It just does not work so well. Humus is the key as Chapter 2 shows.

All of the chemistry, then, will work better if the soil can "breathe" - what a soil scientist calls "gas exchange".

Soil Biology
The bit that is the real works is the soil biology. If the conditions are right, the physical conditions and the chemical conditions, then the micro-organisms can work and will. There are many creatures in the soil but the ones we are interested in are the mycorrhiza. Most of these are beneficial, such as the Penicillins. Some are not and will cause diseases and death in plants. Generally, the good ones are encouraged by the addition of humus (from well-composted organic materials with a good nutrient balance). These fungi will feed the crop

plants at the rate they require it; it is a demand-led system.

What should soil look like? An easy answer is dark and friable. The "dark" colour is desirable because it is an indicator of humus level. The "friable" texture is certainly in part because of the humus but it is also due to good physical and chemical characteristics including fibres and a range of particles from clay to small stones.

How much compost? Well, the evidence is as deep as you like. Remember the Fens? Some started at 10 to 15m deep. The Fens grow good crops and those rich soils do not pollute the dykes and groundwater. There really is no danger to the environment and both vegetables and flowers generally grow well in high organic soils provided they are well drained. (Of course, there are exceptions with particular species adapted to particular soil types.)

High organic matter soils in general warm up earlier in the spring, crops get away faster with less plant disease. There is some evidence that flowers grown on these soils last longer in the bed or cut. Perhaps more important for human survival, there is also some evidence that edible crops have better shelf life, taste better and consumers (wildlife, farm stock and humans) have less disease and live longer$_{39}$.

One word of warning; cultivations accelerate the oxidation of organic matter including the humus$_{13}$. Cultivations also encourage weed seeds to germinate. Where you can, use a mulch and let the worms do the work.

Chapter 9
Time and Soil Bio-Activity

If organic matter, even if it is the form of compost and humus, is put on a soil which previously had low organic matter, then the plants put there will be slow to respond. If that organic matter, same material in the same quantities, were put onto a soil already high in organic matter and humus, the plants respond very quickly.

Repeated here is the figure previously shown as Figure 2.3.

Figure 9.1

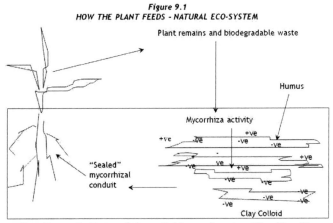

Figure 9.1
HOW THE PLANT FEEDS - NATURAL ECO-SYSTEM

Mycorrhiza are the key to pollution control because they give a "Closed Loop" to recycling both cations <u>and</u> anions.

The plant is not fed directly. It isn't the plant roots that feed on the humus or other organic materials added; it is the soil fungi, the mycorrhiza. It is these mycorrhiza which feed the plant. If there isn't much organic matter in the soil, then there isn't much for the mycorrhiza to live on, so there are less of them. So they can't feed the plants so well. They will, but it may take time to multiply up the mycorrhiza population. Feed them the right way and it may well be quick. If things are close to ideal, it is surprising how rapid the reaction can be. However, soil type, soil temperatures in a particular year and season, moisture content of the soil, the presence of soil disease organisms, including some mycorrhiza which may inhibit the good ones, and many other factors, some of which we know about and some not; all of these may affect the response, and we have little control or even knowledge of some of these. So, the response may be slow, possibly two or three years. However, it does happen. There is real evidence, practical, technical, academic, scientific, common sense, that it really does work - add the organic matter to soils and they work better[36]. Good results come from plenty of humus from well made compost, with a good Nitrogen content and complete in the composting process.

Generally
Some soils need extra inputs. Sands are "hungry" in that the organic matter will easily oxidise and nutrients easily leach out into the groundwater and to the rivers. Good, fertile loams will need it if you cultivate. Clays will hold onto nutrients well but will work better and allow better plant growth if they have plenty of organic matter.

Chapter 10
Human Health and Waste-to-Soil

Public Perceptions of Safety
In general we really do need to move education forward. Far too high a percentage of young people have very limited literacy; so they cannot inform themselves properly. Far too much of the media, in terms of the daily press, is technically illiterate as well, and, more important, the standard of professionalism has slipped into negative sensationalism, rather than an honest balance.

It is easy and common for the public to react to environmental news in a negative way. For the operator, the answer is in a high degree of not just professional integrity based on technical competence, but also in very tactful and active management of publicity.

Government and Supermarket Attitudes to Perceptions
Recycling to land must, sooner or later, result in media and public concerns about the safety of food grown on land where "wastes" were used as fertiliser. The supermarkets will be increasingly touchy about public perceptions on this matter; their presence in the market place depends on managing these matters whether they are justified in science, or not.

There is a simple image of safety which people will find emotionally easy to accept. It is worth going back, again, to the countryside cottage, where the old man lived to be over 100 years old, he grew and ate his own vegetables. It was "organic". The soil is black. It is black because he recycled everything, everything, back to his garden soil.

Is it the same now? Actually, no it isn't because "wastes" are no longer the same. However, our technology is better and we do now have both some records and the technology to deliver real progress; *people who live on food grown on land fertilised with wastes can be healthier and live longer.*

Food Safety
Nature has a remarkable capability; spread something out far enough and give it enough time and it will break anything down. The Titanic isn't rusting at two miles down (there is not enough Oxygen) but it is being *eaten* by bacteria. It will take several hundred years but they will eat it all away.

This raises the legal question of whether "spreading out far enough" is a devious way of using words to avoid "dilution" which is frowned upon by regulators.

Farmers have known for centuries that additions of trace elements help animal health and growth; they used what were commonly called "salt licks". Farmers near the coasts used seaweed as fertiliser and that contained Iodine. "Derbyshire Neck" in humans was an Iodine shortage (Derbyshire is the UK County farthest from the sea.) We have progressed very significantly in understanding how minerals can be used in animal health; much more than in humans but the knowledge is mainly directly transferable. However, there is still much to do on this front. One of the misunderstandings is about "heavy metals" which were originally frowned upon because of excesses in the use of Sewage Sludges. Nearly all these "heavy metals", however, are necessary for health; including Copper, Zinc, Chromium, Arsenic and others including Lithium, Molybdenum, Manganese, Cobalt and many more. For example, if a human has less that 10ppm of Arsenic in their blood stream, they are likely to feel ill and might be dead. Alternatively, if they have 1000ppm, the might feel ill but are more likely to be dead.

So, there is a real question as to how to ensure that the food we eat has enough of the trace elements (we can get them from recycling "wastes") but, at the same time, ensure that the food does not have too much of each or any of these elements.

Human Health
The most difficult question which the technologists in Land Network, the farmers consortium, have begun to wrestle with is not how to monitor these elements in the soil and in the compost of direct spread waste (they already know that), it is how to do this economically with the resources which a small commercially-driven organisation can risk committing. The technology and experience does exist but the trained-up manpower is limited and significant resources are needed. Nevertheless, progress is being made. Safety is certainly about "dilution", and it is also about spreading it out so that nature can cope. Importantly, however, it is also partly about what has *not* been put on the land. Some arable soils have had no organic manures, no animal waste, no compost, just relative pure mineral fertilisers, for 50 years. Ammonium nitrate is just ammonium nitrate - no trace elements. However, in that time, harvested crops will have removed enormous amounts of trace elements.

This, then, defines the big challenge for recycling to land. We have to replace these trace elements[25]. We may not know all the answers but, on evidence, it seems unavoidable that this is a matter of human health and longevity. The most economical way to do this is via "waste". It does also appeal as being "natural". We also have to replace organic matter. "Organic" to a chemist means composed of a large Carbon-based molecule. That means that we can use plastics - not robust plastic sheets (such as polyethylene or "polythene") but, where they exist, waste *liquid* plastics which can be incorporated and spread out onto every particle in the compost mass. We also need fibres in the soil to manage moisture and gas exchange. There will be less flash flooding if we have fibres to hold the surface open. If the land is subject to intensive cultivation, these fibres may deteriorate (by oxidation) too rapidly and synthetic fibres such as carpets (which have been finely shredded) can and will provide that physical function.

Kinsey and Human Health
Sustainability is the key word; in soils, in cropping and in human life.

Back in the 1950's, a soil scientist called Dr William Albrecht[25] had finally concluded how to model the chemistry of farmed soils. He had originally been asked to look at soils which had been broken out of prairie and had yielded remarkably well at first but many went into decline and whatever farmers added in fertiliser, seemed never to recover their yield potential. Albrecht modelled the fertile soils and the declined soils; that was relatively easy. It took years to find a mathematical "bridge" which allowed an operator to calculate how much of what material needed to be added and how long it would take. The bad news was that the calculations were very complex, took a long time and nearly all the world put the work on one side as being commercially unusable. However, the fact is that the model not only worked, it showed something else of staggering importance.

In a study tour shadowing the now world-renowned soils expert Neal Kinsey[14], who took over this work, I was shown a farm where Kinsey had been called in to look at a dairy farm where the grass yields had declined and whatever NPK fertilisers (Nitrogen, phosphorus and Potassium) the farm added, the decline progressed. Kinsey modelled the soil and recommended additions of other plant nutrients, including what farmers call "trace elements". It took years but the grass yields did return to "prairie days" fertility. Something else happened; cows "took to the bull" first time, had more live calves and the calves grew faster. The cows had less disease, stayed fertile for more years and lived longer. The situation was complex and, from this one example, it would be dangerous to conclude too much. However, this is not an isolated incident. There is a significant amount of credible evidence that getting the right mineral elements into the ground not only helps the plant (and the crop), but also helps the animal that feeds on those plants. It is entirely logical to progress this thinking to conclude that a food chain involving waste as fertiliser does positively affect human health and longevity.

Chapter 11
Your Carbon Footprint
Does it Matter Anyway?
The Concept of Carbon Footprint - How to Work Out Yours Within the Garden.
How Your Garden Can Affect What Happens Outside the Garden

Let's face it, the garden is about the best place on earth (sorry about the unintended pun) to be not only "green" but to actually have a low or negative Carbon footprint.

Does it Matter Anyway?
If you doubt all the talk of global warming, then you could do little better than reading the book "The Hot Topic" by David King$_6$, who retired recently as the government's Chief Scientist. The book is a good read. King assembled much of the evidence, presented it in a readable, reasonable and balanced way and yet, despite staggeringly frightening evidence, remained positive and hopeful. There are some who take a very negative view, saying it is already too late. Whatever else is said, we owe it to our children, and indeed ourselves, to give it the best shot we can, starting yesterday.

What we, each individual, can do is look at the idea of "Carbon Footprint". People talk about it very easily, but what does it mean concisely and can you calculate yours easily?

The Concept of Carbon Footprint
Put the simplest way, it is a question of how much Carbon dioxide and other "greenhouse gases" do you leave behind you? While it is possible to do this, there is a difficult job in deciding where to start and where to finish. Also, some gases that might be produced (such

as Methane, for example) are many times more damaging than Carbon dioxide. So, how do you start?

Well, there are some websites, and several are very good. However, they generally relate to households and businesses, not gardens. In fact, in several short surveys on the web, I could not find a garden Carbon Footprint calculator which worked and was free of charge. So here is a start by looking at the principles first and then a real calculation.

PCCS - Photosynthetic Carbon Capture and Storage in Soils
Burning coal with new "clean" technology and CCS (Carbon Capture and Storage such as capturing the Carbon dioxide which can be pumped down into storage in porous rock deep in the ground) has been very much in the news from time to time as the energy debate progresses. However, there is another route to a bigger, cheaper and directly more productive CCS; the plants in a garden can give us Carbon Capture and Storage in Soils. The clever mechanism is Photosynthesis which can take enormous amounts of Carbon dioxide out of the atmosphere (well, that is where our fossilised fuels came from) and pump Oxygen back in.

During crop growth, the chlorophyll in the green leaf of a plant traps the energy in sunlight. This gives the plant the energy to take Carbon dioxide out of the air (via small holes in the leaf known as stomata) and water (via the roots) to form first sugars, followed by carbohydrates, oils and proteins. Indeed, during the Carboniferous Era, starting some 350 million years ago and lasting 60 million years, plants harvested sunlight, took Carbon dioxide out of the air and formed these compounds which eventually formed the black tarry substance which we now drill for as "crude oil".

Nowadays gardening follows the exact same process, using crops to harvest the energy in sunlight. This is known as Photosynthetic Carbon Capture and the Storage is in the Soil. What some farms do is produce oil seed crops and that can be for biodiesel and PPO (Pure Plant Oil) fuels which do exactly what the Carboniferous Era did.

Looking at Figure 11.1, there is some chemistry shorthand that you will be able to follow:

The first line is what happens in words and the second in chemical symbols. The third line remarks that animals burn sugar in their bodies to breathe out Carbon dioxide and it is as well to remember that they use up Oxygen when they do that - and so does incineration.

The rest of the figure balances up the arithmetic to show how many molecules of each material get used in the process and, from that, we can calculate exactly how much material is produced or used - the Figure uses kg but it could be any unit of weight and the figures would still apply.

The bottom line shows what happens when we burn petrol in a car or petrol-driven mower.

Figure 11.1

Figure 11.1
The Basic Equations in Managing Real Sustainability

Energy from the sun
Plants take CO_2 and water ⟶ To make large Carbon molecules

$$CO_2 + H_2O \rightarrow C_6H_{12}O_6$$

⟵ Animals and incineration push this the other way

The balanced chemical equation reads:

$$6\ CO_2 + 6\ H_2O \rightarrow C_6H_{12}O_6 + 6O_2$$

Burning a small Carbon molecule reverses this process:

$$C_3H_8 + 5\ O_2 \rightarrow 3CO_2 + H_2O$$ - plus some energy as heat which we could use for making electricity
Propane Oxygen

Burning a big Carbon molecule would read:

One molecule from Petrol $2C_{36}H_{74}$ + 109 O_2 ⟶ 72 CO_2 + 74 H_2O
Rounded figures   1 tonne + 3.5 tonnes ⟶ 3.2 tonnes + 1.3 tonnes

Never mind the **Carbon dioxide**, where is the **Oxygen** going? The only reversal mechanism we have right now is the **green leaf**.

Forget about the lawn mower for a moment and think about the growing plants in the garden. They are actually reversing global warming! When you grow plants you remove Carbon dioxide from the air, release back Oxygen and, when you make a compost heap, rather than a bonfire, you are locking up quite large amounts of Carbon dioxide. (Note that a compost heap, when it is hot, is producing Carbon dioxide. The micro-organisms do burn Carbon in the organic matter and use some Oxygen to do it. However, these figures are comparatively small.) The figures in the equation above are staggering. As the figures show, the compost heap is definitely a better idea than a fire.

Diagram 11.2 was first drawn up to look at this concept on farm crops but it does apply to the garden, even if it is on a smaller scale. When you grow a green leaf, it harvests sunlight; the garden harvests the energy in sunlight! Just making organic matter using the green leaf of a growing plant takes Carbon dioxide out of the atmosphere and pumps Oxygen back in.

What some farmers can do is grow commercially viable areas and tonnages of oil seeds and produce biofuels[4]. What you can do is find used cooking oil from a chip shop or restaurant and produce your own biofuels. A small kit will cost around £1000 and there are some running costs and you must pay the tax! The good news is that, provided you do not produce more that 5000 litres per annum, you do not need a Permit from the Environment Agency.

What happens in all crop growing is that the chlorophyll in the green leaf of a plant can trap the energy in sunlight which allows the plant to have the energy to take Carbon dioxide out of the air (via small holes in the leaf known as stomata) and water via the roots to form first sugars, then carbohydrates, oils and proteins. During the Carboniferous Era, plants harvested sunlight, took Carbon dioxide out of the air and formed these compounds and, eventually, these formed the black tarry substance which we now drill for as "crude oil". What farming does now, is exactly the same process, they use crops to harvest the energy in sunlight. This is Photosynthetic Carbon Capture and the Storage is in the Soil[18].

Figure11.2
The Basic Sustainable Crop Loop (the way it is often thought of and portrayed). The crop harvests sunlight and turns Carbon dioxide taken in through the leaves, and water taken in through the roots, into large Carbon molecules which we can use a fuel. If we burn the fuel, we get the Carbon dioxide back again. What is often not thought about is that, in taking the Carbon dioxide out of the atmosphere, the plant actually

gives us back the Oxygen. However, it is as well to remember that when we burn the fuel, we burn the same amount of Oxygen as originally released and turn it back into Carbon dioxide.

Figure 11.2
CROPS TO BIOFUELS - The Basic Route

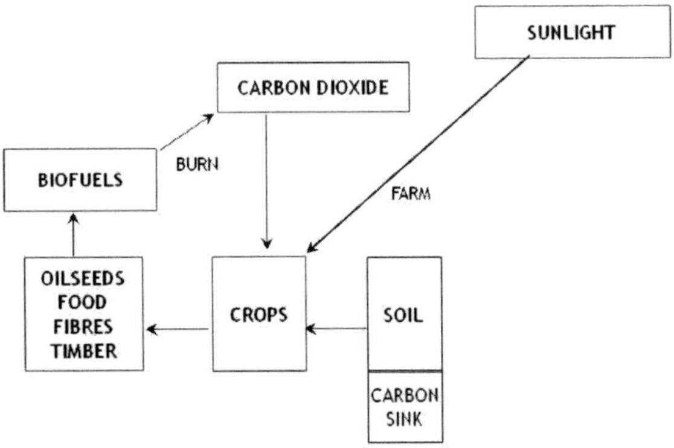

[Diagram: SUNLIGHT → (FARM) → CROPS; CARBON DIOXIDE → CROPS; CROPS → OILSEEDS, FOOD, FIBRES, TIMBER → BIOFUELS; BIOFUELS → (BURN) → CARBON DIOXIDE; SOIL ↔ CROPS; SOIL → CARBON SINK]

The Concept of Carbon Footprint - How to Work Out Yours Within the Garden

Carbon Footprint for Your Garden
If you go to any of a series of websites (just enter "calculating Carbon footprint" into any search engine), you will easily find help in making a reasonable estimate of what figure you are producing from your house itself, your domestic gadgets and your transport including flying. For a couple of adults in a reasonably well-insulated house, run reasonably frugally, you will end up with a figure of maybe 2 to 4 tonnes of Carbon dioxide produced per year, and around 3 to 4 more for your domestic gadgets; say about 5 for your domestic total is not bad. (But, as the school report says; "Could do better".) If you fly much, that figure could easily *double*!

Now, and here is an important point, even a small garden actually consumes Carbon dioxide and locks it up with a slow release. When plants grow and die back and you put the material first into a compost heap, and then dig it into the soil, it locks that Carbon up. It does actually leak back, at a rate of around 35%, if you dig and hoe a lot, down to maybe only 10% a year if you only dig once a year, or just mulch. So, by gardening using compost and with a bit of care, you can "offset" some of your household Carbon dioxide production by locking some up in your garden soil.

Question is; How much?

Try working through the following one step at a time. It is not highly accurate because it is talking typically rather than the actual, scientifically measured, analysis of your garden. However, it will give you a reasonable idea in a way that is not wildly misleading.

We know that one hectare of a really good crop in farming will produce somewhere in the region of 10 to 18 tonnes of dry matter in a year; some crops a bit less, some (such as trees) a bit more. Suppose we take as a guide, 15 tonnes. Reduce this if you have much bare earth. Now, one hectare is 10,000 square metres. So, for one square metre, that works as;

15 tonnes x 1000 kg in a tonne, divided by 10,000 sq m in a hectare = 1.5 kg per sq m in your garden.

Now, most of that dry matter is based on large Carbon molecules. In fact, somewhere over 80% is likely to be Carbon but we will take 60% here to be well within a reasonable limit. The atomic weight of Carbon is 12 and

the atomic weight of Oxygen is 16 and there are two of them, so the molecular weight of Carbon dioxide is 44. That means that if we have 12kg of Carbon in the organic matter in your garden, it will have been made by taking 44kg of Carbon dioxide out of the atmosphere.
Now, go back to that 1.5kg of dry matter produced by each square metre in your garden. Take 1.5, multiply it by 60% (which gives 0.9kg of Carbon), divide by 12 and multiply by 44. That gives a figure of 3.3kg per which is how much Carbon dioxide each square meter an active garden takes out of the atmosphere every year.

Summary;
How many kg of dry matter per sq m?
Multiply that by 60% to give how much Carbon?
Divide by 12 and multiply by 44 to give the weight of Carbon dioxide?

That is how much Carbon dioxide which that sq m of productive ground will take out of the atmosphere every year.

Multiply by the number of sq m of productive ground you have. That is the amount of Carbon dioxide your garden takes out of the atmosphere every year.

How much active garden do you have (and you can count the lawn)? Wow! Now, let's try and make the calculation a little more accurate and related to what you do on your whole patch, including the house.

About half of the dry weight of organic matter produced by a plant is *below* the ground. That is fairly safe from oxidation unless you dig a lot. Above the ground is more vulnerable to loss; you may take vegetables off, cut flowers, prune bushes and trees. If you compost most (by weight) of that, the organic matter is, again, fairly safe but you will lose a little in the compost operation. You can't count any of the dry matter if you burn it. Compost will usually lose a quarter to a half of the weight put into the process. Most of that is water which is boiled off. However, the heat in a compost heap is produced by the micro-organisms using the Carbon for energy production - just as your body "burns" sugar during exercise. That energy loss, however, is usually quite small, only a few single figure percent. If you add to the compost heap by putting in imported organic wastes (not the waste from vegetables you grew in the same garden) but vegetable waste you bought from outside the garden, or some horse manure or Farm Yard Manure (FYM) - or whatever, then you can add that to your Carbon "sink" calculation. (The "Carbon sink" is the Carbon you have put into the soil bank.) It is the same sort of calculation as follows.

Every kg of imported material will have much water in it. If it is just lettuce leaves, over 90%. If it is mixed vegetables, probably about 65 to 80%. If it is FYM, probably 65 to 70%. Horse manure; may be as little as 30%. You can either guess at this or weigh a sample and then put it in the oven at around 110°C for at least an hour or until the weight stabilises. Weigh again. Now, for every kg of dry matter, you can use the 60% figure for the Carbon and then the same calculation;

1kg dry matter x 60% = 0.6kg

0.6 divide by 12 and multiply by 44 equals 2.2kg of Carbon dioxide taken out of the atmosphere to make that organic matter.
Suppose, over a year, you import 1 tonne of other materials with a total dry matter content of 500kg. That

would have taken, somewhere down the line, at least 1100kg or 1.1 tonnes of Carbon dioxide out of the atmosphere. Now, you can only charge that to your Carbon offset if it would otherwise have been burned. Might it be? Well the government is pressing on with what it calls Energy from Waste (or EfW for short). The process does yield some energy and that is instead of burning fossilised fuel but it is still burning and still returning the Carbon to the atmosphere as Carbon dioxide. (Nothing is simple, is it?)

Unfortunately, most gardeners are also adding Carbon dioxide by another route; crop protection chemicals and mineral fertilisers. Both use energy and/or petroleum products to make them. That energy production will produce Carbon dioxide.

Because the amount of plant protection chemicals used in a garden is small in terms of kg, and the fact that this calculation here is a useful guide rather than a highly accurate scientific study, it is not unreasonable to disregard that from the point of view of this calculation. However, there is a small figure and it should be avoided as far as possible and not forgotten where you do use such materials.

Fertilisers, if they are mineral or "artificial", are a different story. The worst is Nitrogen. It is made by passing air through a large, continuous electric arc - usually 2 metres in diameter. It takes enormous electrical power. Globally, the electricity is, of course, mainly made by burning fossilised fuel which produces Carbon dioxide. These production plants do vary in their efficiency in using electricity and modern facilities may recover some Carbon dioxide gas. Nevertheless, there is always a net output and this may be 5 to 10 tonnes of

Carbon dioxide per tonne of Nitrogen fertiliser. So, for every 1kg of mineral Nitrogen fertiliser you use, fine yourself 10kg of Carbon dioxide put onto your Carbon footprint. If you are using a mixed phosphate and Potassium (but still mineral fertiliser) you could halve this figure. The important thing here is that the production and transport of mineral fertilisers round the world does consume energy. These materials help us feed the world and they are necessary to do that, not everybody can be fed without them. Nevertheless, their production and transport unavoidably produces Carbon dioxide and if you use them, deduct a suitable figure off your garden footprint. As a guide, not less than 5kg of Carbon dioxide for every kg of mineral fertiliser you use.

Now, to summarise;
1. If you have an idea of how many square metres of active plant growing you have (90% green for 6 months of the year), you can multiply that by 3.3 (actually, it is probably over 4) to give the lock up of Carbon dioxide which your garden has removed from the atmosphere. If you have 33.3 square metres of really active growth, then you just might be taking out from the air what your house put in.
2. You can push that figure up by adding in an amount for "imports" of organic matter at the rate of about 1.1kg of Carbon dioxide for every 1kg of fresh weight import.
3. You must knock off some from that total to account for imports of materials which used energy for their production - chemicals and fertilisers at a rate of 5kg of Carbon dioxide for every 1kg of these materials you use.

Clearly, you would need quite a big garden, and you would have to work quite hard at all of this to offset all the Carbon dioxide you might produce in ordinary living but you can make a hole in your net total.

How Your Garden can Affect what Happens Outside the Garden
Do the calculation.

Pass on the word - we can all make a difference.

Do it and tell your local newspaper - be the first locally to do that.

Write to BBC Country File.

Tell your MP.

Urge your local Council to consider recycling to land - farm or garden both help.

Organise other gardeners and ask your local Council recycling officer for his/her assistance.

Stop. Think. Do it now.

Alternatively, just quietly do your bit and feel good about it.

APPENDIX
References and Further Reading

1. *Soil Husbandry*, Batey T, Soil and Land Use Consultants Ltd, 1988 p52.
2. *How the Closed Loop Delivers True Sustainability*, Butterworth B, EnAgri, Issue 30, September 2008, p27.
3. *Waste in the Next Millennium*, Butterworth B, Resource, American Society of Agricultural and Biological Engineers, July 1988.
4. *Biofuels from Wastes*, Butterworth B, ReFocus, May/June 2006.
5. *Secrets of the Soil*, Tompkins P and Boyd C (1989). Harpet and Row, New York.
6. *The Hot Topic*, King D, Bloomsbury 2008
7. *Nitrate Nonsense*, Butterworth B, Landwards, Institution of Agricultural Engineers, Early Summer 2002.
8. *An Analysis of Composting as an Environmental Remediation Technology*, US EPA, EPA530-B-98-001 March 1998.
9. *Reversal Global Warming*, Butterworth B, ReFocus, September/October 2006.
10. Many researchers have looked at micro-organism numbers but it is those in the USA who have published most work (as any web search will show).
11. *Clamping Down of Compost*, Butterworth B, Resource, American Society of Agricultural and Biological Engineers, April 2006.
12. *Plant Litter, Decomposition, Humus Formation, Carbon Sequestration*, Ed Berg B, et al, Springer 2003. This is just one of the many hundreds of references to organic matter degradation and formation of humus, in the academic literature and on the web.

13. This particular research is part of Soil Resource Management, an ARS National Program (#202) described on the World Wide Web at http://www.nps.ars.usda.gov. Wright Sara F and Nichols Kristine A, are with the USDAARS Sustainable Agriculture Systems Laboratory, Beltsville, Maryland, USA.
14. *Hands on Agronomy*, Kinsey N, Acres USA, 1999.
15. *Biogeochemical Characterisation of Metalliferous Wastes and Potential of Arbuscular Mycorrhizae in their Phytoremediation*, Chaudhry T M, School of Science, Food and Horticulture, College of Science, Technology and Environment, University of Western Sydney, Australia.
16. *The Contribution of Mycorrhizal Fungi in Sustainable Maintenance of Plant Health and Fertility*, Jeffries P, Biology and Fertility of Soils, 2003, vol 37, p1-16.
17. *Signalling in the Arbuscular Mycorrhizal Symbiosis*, Harrison M J, Annual Review of Microbiology, 2005; 59 p19-42.
18. *The Design of a Pesticide and Washdown Facility*, Rose S C et al, British Crop Protection Council Symposium, November 2001.
19. *Capturing the Future for Sustainable Environments*, Butterworth B, Far Eastern Agriculture, May/June 2008, p12-13.
20. *The Straw Manual*, Butterworth B, Spon 1986.
21. DTI unpublished reports by Land Network International Ltd under the Enterprise Initiative Programme.
22. *Development of Compost Maturity and Actinobacteria Populations During Full-Scale Composting of Organic Household Waste*, Sterger K, Journal of Applied Microbiology, 2007.

23. *Microbiological Aspects of Biowaste During Composting in a Monitored Compost Bin*, Ryckeboer J, Journal of Applied Microbiology, 2003, 94 (1).
24. ICI Plant Protection, as was, pursued their market for Gramoxone with both commercial and academic enthusiasm in the 1980's with many publications and support for trials, including by universities.
25. *The Albrecht Papers*, Albrecht W A, Charles Walters Books, 1919 to 1970.
26. *Managing the Soil Rumen and Ion Exchange*, Butterworth B, Arable Farming, 9 September 2000.
27. There are many references in many places including on the web, for example at Washington State University, Cornell University and others.
28. Gordon Spoor, was a lecturer and researcher at the National College of Agricultural Engineering, Silsoe, UK, for more than 20 years in the 1980's and 90's. He was acknowledged as a world expert on soil strengths and published many papers on the subject.
29. *Upland Britain*, Millward R and Robinson A, David and Charles (Publishers) Ltd, 1980.
30. *Materials and Energy from Municipal Waste*, Office of Technology Assessment, United States, Diane Publishing 1979.
31. *Microbial Activity in Blocking Composts. 3. Degradation of Ureaformaldehyde*, Turner C P and Carlile W R, ISHS Acta Horticulturae 150, International Symposium of Substrates in Horticulture other than Soils in Situ.
32. There is much research on composting and the processes involved done by several of the USA universities.
33. *Black is the New Green*, Sequestration, Vol 442, P624-626, 10 August 2006, Nature Publishing Group.
34. *Our Good Earth*, Mann C C, National Geographic, September 2008, p80-107.

35. Evans T, personal communications with Land Network.
36. *Waste in the Next Millennium*, Butterworth B, Resource, American Society of Agricultural and Biological Engineers, July 1998, p11-12.
37. *Mycorrhizal Fungi Influence Competition in a Wheat-Ryegrass Association Treated with Herbicide Diclifop*, Rejon A *et al*, Applied Ecology, 1997, 7, p51-57.
38. *With or Without MDF?* Butterworth B, Far Eastern Agriculture Sept/Oct 2003 (also Sept/Oct 1987).
39. *The Soil Health*, Howard A, Scjhocken, New York 1947.
40. *The Humanure Handbook*, Jenkins J, 2005 Third Edition, Joseph Jenkins Inc, Pen USA.
41. Research by Land Network International Ltd, published later in Biofpr, December 2008.

About the Author

Bill Butterworth has moved house 16 times in his marriage with Christine, his wife of over 40 years. Apart from initial landscaping, he claims not to have "dug a garden" in the last 30 years. He is a big fan of working *with* nature and enabling it to do as much work as possible.

Bill's first degree was taken at Reading University in agricultural science back in the early 1960's. He was a lecturer for 13 years in Essex at Writtle College, where more than half the students were horticultural and was, for 21 years, a Moderator at the School of Horticulture at the Royal Botanic Gardens, Kew. He says that he probably was the first person to publish a paper on how the mechanisms of the "Closed Loop" really work but also says that what really matters is knowing how they work "because if you don't, you cannot possibly manage them".

Bill Butterworth, BSc Hons (AgSci), CEnv, CM, FIAgrE, FCILT, FCIT, FCIM, MCIoJ, MASABE, MCInstWM, is the Managing Director of Land Network International Ltd (www.landnetwork.co.uk) which provides a world class involvement in recycling wastes to land. He is the architect of the "Reverse Franchise" consortium used in Land Network, the farmers' waste-to-land recycling project operating nationally in the UK and he has specialist knowledge in recycling controlled wastes and in pollution control. In the past 17 years to date he has worked exclusively on recycling waste to land, publishing papers and books on the "Closed Loop" mechanisms, pollution control, on Carbon Capture using agricultural crops fertilised with wastes, and on integrated biofuel and food production from wastes.

Printed in the United Kingdom by
Lightning Source UK Ltd., Milton Keynes
141769UK00001B/30/P